Seeing Double

Seeing Double

Shared Identities in Physics, Philosophy, and Literature

Peter Pesic

The MIT Press
Cambridge, Massachusetts
London, England

This book was set in Sabon by Achorn Graphic Services, Inc. on the Miles System and was printed and bound in the United States of America.

Portions of this book appeared originally in the following journals, which have kindly given permission for the appearance of the material here: *The St. John's Review* (© St. John's College, 1989), *Literature and Theology* (© Oxford University Press, 1994), *Studia Spinozana* (© 1996), *Philosophy Now* (© Peter Pesic 2000), *American Scientist* (© Peter Pesic 2002).

Permission for use of the figures has been kindly given by the following: The Burndy Library, Dibner Institute for the History of Science and Technology (figures 3.1, 6.1, 6.2); courtesy of the Archives, California Institute of Technology (figure 8.5); courtesy of IBM Research, Almaden Research Center (figure 8.6); courtesy of David Wineland, NIST (figure 8.7). I am grateful to Natalie Zemon Davis (figure 4.1), Don Eigler (figure 8.6), and David Wineland (figure 8.7) for generously making these images available. I would also like to thank Anne Battis (Burndy Library), Anthony J. Leggett, Alexei Pesic, Andrei Pesic, and Ssu Weng for their help with the figures.

Library of Congress Cataloging-in-Publication Data

Pesic, Peter.
  Seeing double : shared identities in physics, philosophy, and literature / Peter Pesic.
    p. cm.
  Includes bibliographical references and index.
  ISBN 0-262-16205-9 (hc. : alk. paper)
  1. Identity (Psychology). 2. Individuality. 3. Psychology and literature. 4. Psychology and philosophy. 5. Philosophy and science. I. Title.
  BF697 .P47 2002
  111'.82—dc21
                                                          2001041012

10 9 8 7 6 5 4 3 2

For S., A., and A.,
beloved individuals

Contents

Prologue

Imagine a winter evening with snowflakes falling. Each snowflake is unique, an irreplaceable individual, or so one has heard since childhood. But is it really true? This book will show that each snowflake is composed of beings that are indistinguishable in every observable respect. Electrons and all other species of elementary particles exhibit no individuality. The members of each species are identical to a degree that is without parallel in the domain of ordinary human experience. As a consequence, they show a strange interdependence that manifests itself in the peculiar phenomena described by quantum theory. All aspects of chemistry depend crucially on this loss of individuality, as does practically every branch of physics.

This book will often traverse the boundary between physics and philosophy. As presently conceived, these are separate fields, each rather technical and circumscribed. Yet the words "scientist" and "physicist" were only coined in the nineteenth century. Isaac Newton, Michael Faraday, and James Clerk Maxwell called themselves "natural philosophers," using the ancient word for the inquiry into nature as part of the "love of wisdom"—in Greek, *philosophia*. The questions of natural philosophy deserve the widest possible audience not only for their social and political implications but even more for their sheer beauty and fascination. Here, popularization is perilous. Thoughtful readers want to confront the deep questions; they are not satisfied with dazzling but unreflective displays of scientific legerdemain. The life of Socrates offers a different example: real thought is a conversation among equals about the most important questions. Respect for the reader's intelligence calls for the fullest sharing of these questions, inviting further thought and conversation. This is not merely a polite fiction to keep "laypersons" from feeling inferior. In truth, all of us have only begun to ask the question of identity. Its strange and mysterious story calls for much further reflection.

To tell this story, I will reach far back into the earliest human meditations on individuality and identity. I hope their connection with modern developments will gradually become apparent and compelling. After all, the question about the individuality of electrons goes back to the most fundamental human confrontations with identity, which prepared us to formulate this question in the first place. The connection or apartness of individuals is perhaps the central question of human life: What, exactly, is *my* individuality? To what degree is it incommunicable or unique? To what degree can it be shared, and how? If fundamental particles have lost their individuality, one wonders what this may imply for us. Yet this human questioning must respect the aspects of nature that go beyond the human.

My goal is to bring these questions to life and to open them to broad consideration, so that they might begin a thoughtful conversation among many people, not just those already versed in physics or philosophy. To do so, I have omitted many important aspects, and I have not tried to offer a complete account of all the issues at stake. Because of this, I have not done justice to many worthy thinkers, ancient and modern, who have spoken to these questions, and I apologize to those I have neglected. I have given only basic elements of complex and difficult arguments, but I have tried to be faithful to their central issues. What is "elementary" may finally be what is most important, for everything that is more "advanced" hangs on it. We should reconsider these most basic ideas freshly, or we will find ourselves accepting them blindly. This applies to experts as well as to beginners, and I hope both will find something here. Accordingly, I have chosen a few important passages that raise the question of individuality with particular urgency. To intensify the dialogue, I offer not only questions but also a view that I hope will be helpful and provocative. I will argue that the crux of modern quantum theory is precisely its clash with our ordinary concept of individuality. This represents a significant departure from the way this theory has usually been understood. What is bizarre about quantum theory may then become more intelligible as we reconsider what we mean by individuality and identity in ordinary experience, the better to gauge what is strange and new. To do so takes us back to the sources, the earliest writings that meditate on these ideas.

The ancient sources are not naïve, though. As will shortly emerge, Homer already saw both of the main possibilities, seminal for all later conceptions: individuality is either unique or it is somehow interchangeable. To be sure, he more often considers the individuality of persons than of things. Yet throughout this book personal individuality will haunt the individuality of things. From one point of view, persons and things must never be confused; ethically, things may be used, but persons should not. Still, we naturally take personhood as a touchstone for all individuality, if only to compare our personal uniqueness with the condition of mere things. Throughout, I will use this touchstone to seek the deeper possibilities suggested by the extremes of persons and things, moving freely between them to probe their similarities and differences.

In this quest, examples from literature are often helpful, and not merely as analogies, for they evoke potent possibilities whose vividness can open the mind. Our imaginations need this help, especially when the loss of individuality takes us far from the familiar terrain of everyday experience. Even if my examples may seem at first distant from physics, each aims to illuminate an aspect of individuality faithfully. Throughout, I have not presumed that the ideas presented are familiar to the reader, but have tried to bring them to life anew. Rather than burdening the text with technical details or scholarly references, I have put these in the back, where the reader can also find suggestions for further exploration.

As literature offers help to the imagination, philosophy aims to make our speech and thinking more precise and thus more just. I have selected only a very few instances from the vast diversity of philosophical approaches to the problems of identity. As they have explored these approaches, philosophers have gradually found more precise terms that avoid many misunderstandings, and I will try to use these terms. Our ordinary word "identity" covers several crucial concepts that should be clarified, especially individuality, self-sameness, and distinguishability. I will take the word *individuality* to indicate what makes an individual be individual, rather than simply a member of a species or an instance of some universal quality. Thus Homer's character Hector, the Trojan champion, is an instance of the universals "human being," "man," and also "hero," among others. As an individual, though, Hector is *not* merely an instance of "Hectorness": he is unique. In these terms, I will take an individual to mean a being that is "noninstantiable," not an instance of some universal or general quality, but a unique being.

As an individual, Hector remains Hector in some respect wherever or whenever we encounter him, whether he is alive or dead, real or imaginary. Following the usage of philosophers, I will call this continuing self-sameness his *identity,* meaning the way Hector remains essentially himself in all times and places, even though he may also change in important respects. Note that identity is not the same as individuality, which is more inclusive. Yet the name God disclosed to Moses, *I am that I am,* is the deepest statement of identity.

Hector is also different from everyone else in certain definite ways (his looks, his actions) and so is *distinguishable* or *discernible* from all other individuals. These terms may refer to our knowledge of Hector, not necessarily to what he is in himself. Yet later, in the quantum realm, *indistinguishable* will often mean that no observer whatever, however keen-sighted, could discern any difference. Again, distinguishability is not the same as individuality or identity. Since it is important not to ignore observable qualities, I will consider individuality as extending not only to persons and things but also to their features and characteristics, however they are related. Though one can use the general word "brown," the precise colors of my sons' eyes are as individual as they are; in contrast, neither an electron nor its charge are individual. Here and throughout, much rests on the thoughtful application of these terms.

The quest for the true nature of individuality will take us from modern physics to the dark passages where Franz Kafka sought the nexus joining the human and the nonhuman. It begins with encounters on the windy fields before Troy and with a dark ship racing away from the Cretan labyrinth. Such is the winding maze in which the human spirit seeks its center, wrestling with nature for its secret.

1

Commodity and Sacrament

Early in Homer's *Iliad*, Hector returns to Troy during a break in the fighting; he is seeking his wife and young son. Homer describes the shining helmet that Hector wears as if it were the visible emblem of his heroic nature, so that "shining Hector" gleams with the same brilliance as his helmet. At last he finds his wife and confides to her his fatal prophecy: the day will come when sacred Troy must die. He reaches out for his son, but the boy recoils, screaming at the sight of his own father, "terrified by the flashing bronze, the horsehair crest, the great ridge of the helmet nodding, bristling terror, as it seemed to him." Hector laughs and quickly lifts the helmet from his head and sets it on the ground so as not to frighten the boy. Yet the helmet remains brightly shining even on the ground, as if it glowed with an inner light all its own.

Hector's helmet retains the brilliance of its wearer, even when he takes it off. Here, as often in Homer, inanimate objects, like persons, seem to have an individuality that keeps shining forth, as if to say: Here I am. This individual radiance retains its identity, its self-sameness, even in greatly changed circumstances. The helmet keeps its mysterious glow even when its master is absent, as uniquely radiant as Hector's own features. Later in the story, Hector remains himself even when he runs in uncontrollable terror from Achilles. Intense individuality characterizes each hero and reflects his unique destiny.

Yet Homer's sense of the importance of individuality is shadowed by the grimness of fate. This is illuminated by an encounter on the battlefield, where a warrior would challenge his opponent, hidden behind armor, to reveal his name. As the Greek hero Diomedes prepares to fight an unknown Trojan, he asks him, "Who are you, my friend?—another born to die?" Diomedes wants to knows his opponent's identity so he can avoid fighting an immortal—he has just wounded Aphrodite and aroused the anger of the gods—and so that he can gauge the glory he will gain by slaying his opponent. Glaucus answers with strange detachment, as if that glory were bitter to him:

"High-hearted son of Tydeus, why ask about my birth?
Like the generations of leaves, the lives of mortal men.
Now the wind scatters the old leaves across the earth,
now the living timber bursts with the new buds
and spring comes round again. And so with men:
as one generation comes to life, another dies away."

If men really are like leaves, no true individuality sets one apart from another, and nothing special can be gained by trying to pick out one leaf from another. The endless cycle seems to sweep aside the distinctions of honor and parentage, or at least seems to make those distinctions come to nothing. In this dark view, individuality is effaced by omnipresent death.

Yet Glaucus knows Diomedes by name and knows that he is the "high-hearted son of Tydeus." Despite his bleak words, Glaucus goes on to tell his own story and lineage; he expresses pride in his royal lineage and determination not to shame his parents. Not all those leaves are indistinguishable, it seems, for some will flame in brilliant colors as they fall, showing in their end something that sets them apart from the rest. After hearing this story, Diomedes recognizes Glaucus as a guest-friend, a hereditary accolade based on the hospitality that Diomedes' father once granted to Glaucus's father. Not only is Glaucus spared a mortal combat, but Diomedes removes him from the ranks of his enemies and offers to trade armor as a sign of their special relationship. In this case, Glaucus's individuality has not given him glory in death, but instead saved his life.

Yet, despite their fortunate recognition, Homer tells us with unusual directness that, by accepting Diomedes' offer to exchange armor, "Zeus stole Glaucus's wits away. He traded his gold armor for bronze with Diomedes, the worth of a hundred oxen just for nine." It turns out that Glaucus does not understand the true value of exchange, after all; his description of the interchangeable leaves shows that he does not fathom the extent to which one leaf, or one man, cannot be substituted for another. To punish Glaucus's lack of discernment, Zeus befuddles his wits. Showing his own awareness of what is at stake, Homer carefully names the price of armor not just in terms of honor but in the practical currency of oxen, the common medium of exchange.

Such exchanges slowly developed into the use of coined money, which is roughly contemporaneous with the first written versions of Homer's epics. In Greece, as in other early civilizations, records of accounts and inventories are among the first written documents to be found, and frequently are the earliest. Indeed, the beginnings of writing may be related to economic life of a certain complexity, for religious and political matters may not require written expression in the way that business does. In Greece, important public accounts were written, engraved in stone and placed on public view, for the leaders of the city were accountable to their fellow citizens as well as to the gods. Every officeholder had to present written records to auditors at the end of his term or else he could not go abroad, dedicate a sacred offering, or make a will.

Even before money became the common medium of exchange, the individual could be a commodity, bought and sold. The issue of exchangeability is crucial throughout the *Iliad*; it tests when one individual can be substituted for another. At the beginning of the story, Achilles refused to accept any substitute when his cherished war-prize, the girl Briseis, was unjustly taken away from him by his commander. No amount of gold or other women would compensate for her loss; any suggestion of barter or substitution was despicable. However, Achilles later allows his beloved friend Patroclus to go into battle wearing his own armor, as if that allowed one friend to act as a substitute for the other. When Hector kills Patroclus and puts that armor on himself, the exchange of identities turns another way: the Greeks see the armor of Achilles fighting against them, not at their side. After Achilles avenges his friend, he finally allows the Trojans to ransom Hector's body, an exchange he had previously rejected with contempt. Even the greatest and most singular of men finally does not refuse all exchange.

Thus far, individuality has been open and manifest, so that the interchangeability of different individuals remains a matter of weighing the relative values of known quantities: should this armor be exchanged for that? Glaucus should have paid attention to plain facts: bronze is not equal in value to gold. In his dark mood, he did not sufficiently heed the special value of the armor his family had given him. In this case, the different objects were readily distinguishable. Homer also presents far more difficult possibilities. His version of the story of Odysseus puts into question the manifest openness that distinguishes Hector and his helmet. The wiliest of all the heroes, Odysseus is also the most deceptive, leading a night raid into enemy territory by pretending to be a Trojan. Sometimes identity is deeply concealed, as when Odysseus disguises his identity in order to return home safely. Despite his skill at deception, Odysseus's self-concealment is deeply problematic. For instance, when he first landed on the island of the Cyclops, Odysseus concealed his identity under the devious name "No-man." After he succeeds in blinding the Cyclops, his men beg him to leave the island incognito but Odysseus shouts out his real name, as if his identity was burning to disclose itself. Learning this, the Cyclops's divine father curses and hounds Odysseus.

During his succeeding journeys, Odysseus guards his identity closely, disclosing it only when unavoidable. He stays seven years with the goddess Calypso, who offers him the gift of immortality, but at the price of remaining in hiding with her forever. Though he replies to her diplomatically, he is not tempted. The prospect of living for an eternity while stripped of his true identity, separated from his wife and son, reduces him to misery, gazing out at the imprisoning sea. It is in this way, weeping bitterly, that Homer allows us first to see him.

Odysseus does not wish to remain hidden, but he cannot return home without concealing himself. He concocts false names and stories to protect his true identity. His deception runs deep; when Odysseus at last awakes on his native shore, he does not recognize it at all and curses fate at the very moment he should bless it. Though his journey had been perilous, the greatest dangers of all await him in what should be the safest haven, his own home. To regain his true place, he must summon his highest powers of deception.

Odysseus must not only deceive the suitors in his disguise as a beggar; he must control his overwhelming anger and keep it in reserve. Though always guided by his true self, he must conceal it so that no one might divine the truth before the chosen moment. Yet, he remains the same underneath. His old dog, Argos, who has waited for him for twenty years and is now old and weak, still recognizes him immediately. The dog dies wagging his tail, before he can betray his recognition to other, unfriendly eyes. Odysseus gradually decides to disclose himself, first to his faithful swineherd and his son Telemachus, but only after testing whether they can keep his secret. It remains mysterious at what point Penelope recognizes him; there are signs that she recognizes him, unbidden, long before all the others do, but keeps that recognition hidden even from Odysseus. If so, she keeps her secret even after the climactic fight with the suitors, holding herself back from the victorious Odysseus. She holds a final test in reserve; she requires from him the secret of their marriage bed. Penelope emerges as his true wife not merely through fidelity but through her ability, perhaps even exceeding his, to remain hidden, to test, to wait for the crucial moment in readiness.

This final test reveals the identity that Odysseus and Penelope share. Still holding herself aloof from the stranger, Penelope proposes to have her bed moved out of the chamber for him. This enrages Odysseus, for he had designed his house so that a living olive tree filled his bedroom. Odysseus trimmed the tree to make their marriage bed; he is furious to think that someone might have chopped the tree down. Only at this point does Penelope break down and call him by name, "recognizing the strong clear signs Odysseus offered."

This sign must be secret to retain its potency. Such secrets must be shared at the crucial moment, but they must be carefully guarded from prying eyes. Those in the inner circle who await Odysseus struggle to maintain their faith, biding their time while others try to usurp his identity by taking his wife and home. The agony of waiting at home may be harder than the lot of the wanderer; Odysseus's mother succumbs to despair and dies, while his father withdraws from the city to nurse his misery. Those who remain behind live out Rilke's bitter truth: "Who speaks of victories? To endure is everything." Telemachus must somehow find the strength and discernment needed to recognize a father he has never known and whose survival he doubts. He must seek from the shadows of his father's glory the beginnings of his real identity, for whose secrets he must go in quest.

Homer makes clear that these secrets emerge only through the intervention of a god. Odysseus's protector, Athena, the goddess of wisdom, rouses his son from boyish reveries at home and nerves him to go abroad to seek his missing father. To do this, she takes on the form and voice of a family friend. By their ability to take upon themselves all sorts of appearances, the gods show their familiarity with the secrets of identity. This is the deepest reason that underlies the ancient rules of Greek hospitality, which require that a guest be treated with utmost respect, and that he not be pressed to disclose his name before enjoying a ceremonious welcome and food. The *xenos* is a stranger as well as a guest, and may be an honest traveler, a brigand, or, most dangerously, a god. Many myths recount the visits of gods among men and detail the terrible vengeance the gods exact if the divine stranger is not properly honored.

Since a stranger might be a divine visitor in disguise, one must view his or her outward appearances with circumspection. Some tell-tale sign may give away the guest's true identity, for good or for ill. However, no mere human guile is sufficient here: Homer makes clear that the gods let slip such signs only when they will and to whom they choose. The meaning is clear: only a god can truly disclose himself to a mortal, just as the gods are immediately recognized by each other, regardless of disguises.

Identity is imprinted beyond human auspices, indelibly; the gods inhabit the larger domain in which individuality is forged and they can reveal its signs, though its mysteries lie in the dim realm of destiny. Here individuality is a *sacrament:* the union of something visible or perceptible—a name, a face—with something invisible and numinous: the hidden, indelible mark that sets Odysseus apart from all others, regardless of disguise or the change of time and age. It goes far beyond visible, distinguishable signs to a "primitive thisness," as later philosophers called it, meaning no particular quality (which might be changeable and could disappear), but an altogether different sense in which *this* person or thing is not *that.* Such deep individuality endures even in the shadowy realm of Hades; what is eternal must be of the gods.

The mark of identity has a divine indelibility that can only be witnessed, not comprehended. Telemachus knows somehow that he must follow his friend—the deity in disguise—with careful steps. If the god is gracious, Telemachus will find his father and also himself. The mystery of individuality may be a sacrament that seeks communion.

2

The Ship of Theseus

The Minotaur lay dead in the labyrinth. For many years, he had devoured seven youths and seven maidens sent each year by the Athenians as tribute to the Cretan king, Minos. Then Theseus, the son of the Athenian king, voluntarily sailed with the young people. With Ariadne's thread to guide him, he killed the Minotaur and saved his fellow prisoners. On his return journey, Theseus forgot to hoist white sails, instead of the black sails of the outbound voyage; his father, seeing what he took to be a funeral ship approaching, presumed his son was dead and threw himself into the sea.

The Athenians, for their part, had made a vow that, if the young people were spared, each year they would send that same ship to Apollo's sacred island of Delos with thank-offerings and sacrifices. Upon his safe return, Theseus kept this vow and began sending the ship to Delos. The gratitude of the Athenians moved them to continue this practice even after the death of Theseus. After a number of journeys, the original ship needed repair. Out of reverence, it was decided that every part replaced on the ship should be indistinguishable from the original, so that the ship would always be the same. Over time, gradually every component of the ship was thus replaced, so that the ship showed no outward change in appearance.

At this point, the custom attracted the attention of thoughtful people. Was the ship really the same or not? Although it did look in every respect unchanged, everyone knew that no part of the ship was identical with any part of the original, in the sense of having the same "primitive thisness." But then how could the ship be said to be the same, if it was indeed different piece by piece? Could its continuing identity, its essential self-sameness, be only the property of its outward appearance, without regard to its material constitution? This became an important question for ancient Greek thinkers. Those who were called the atomists thought that everything was composed of unchanging, indivisible atoms. Their arguments rested partly on the common experience that one can divide an object into many smaller parts and partly on the assumption that matter cannot be divided into nothingness. The palpable reality of material bodies rested on the ultimate integrity of the smallest constituents, the atoms.

Leucippus and Democritus, the earliest thinkers to advance these ideas, could carry out such dissections of matter only in their imaginations. Since atoms are too small for human perception, explanations in terms of atoms must reach beyond the senses. Other Greek thinkers found this peculiar, if not ridiculous. As a young man, Socrates turned away from this kind of natural philosophy because it "explained" everything without ever asking *why* the atoms form the configurations they do and what necessitates the forces of nature. Plato depicts Socrates making fun of the fashionable use of atoms to explain everything while understanding nothing. Plato's student Aristotle added that it is laughable to "explain" the evident in terms of the obscure, since a real explanation takes the obscure and makes it evident. His joke implies that the atomists could not tell the difference.

In contrast to this emphasis on matter, Plato held that *form* was the ultimate reality. Here "form" does not mean simply the visible form, but the "look" of something, that which makes it a coherent whole, not a heap of unconnected aspects. Plato's touchstone is mathematics; all trios of objects share Threeness, without which we never would grasp what is common to three ships and three gnats. True reality lies in those invisible forms, of which individual examples are only shadows. We recognize the ship of Theseus by comparing it with the form we already know, its original archetype, as the real thing. Here *real* denotes something that not merely exists but is what it seems to be, something genuine, as in "a real antique" or "a real man." What is *real* has a kind of inner integrity; the ship of Theseus is real if it maintains its original, authentic form. Such a form, however, is not an individual but a universal that transcends any specific material ship. By locating ultimate reality in the form, Plato denied that any individual is real. According to Plato, since what we call existence is bound up with individuality, there must be a transcendent realm beyond both existence and individuality. What we call existence is only the shadow of something far greater.

Aristotle tried to balance the changing world of appearance with the changeless realm of form. Accordingly, he judged that the ship of Theseus did not remain the same in *matter,* though it was the same in *form.* This elegant formulation invites the question: Can one really divide the individuality of the ship into two disparate aspects? For his part, Aristotle did not want to separate form from matter, unlike Plato. Aristotle maintained that one begins with *individuals* that really exist in the world, and which only afterwards are grouped into species or genera. The individuals are primary. Yet we understand each individual in terms of its form, its membership in a certain species or genus of beings.

The atomists reached beyond immediate human experience. In place of commonsense clarity, they turned to the logical satisfaction that atomic accounts could give. The combinations of a few species of atoms might explain the myriad variations of matter. Democritus seems to have held that there were an infinite number of types of atoms, each type differing from every other. Since his surviving writings are fragmentary, it is not clear exactly what he thought on this point; the continuous diversity of atoms was a way of taking into account the vast diversity of visible bodies. Yet this is a problematic conception that raises many questions, despite its explanatory power.

The atomic theory began by accounting for the individuality of each visible body by appealing to the collection of atoms constituting it. Thus the apparent individuality of a body comes from the arrangement of atoms in it, for atoms are common the world over. But if, as Democritus seems to have thought, each atom is distinguishable from all others, we are back where we started. Instead of talking about the individuality of the body, we have merely shifted to specifying the individuality of each atom in that body. Furthermore, if the individual atoms differ in shape, one immediately wonders: "What makes this atom stick out *here,* and that one *there?*" It is inescapable to think of some still smaller constituents that give each individual atom its peculiar shape. Those subconstituents would be the atoms' atoms, as it were, and the same argument will circle around again unless one finally insists that the *real,* ultimate atoms are generic components lacking differences in structure that would tempt endless appeal to sub-subatoms.

Although in later times the fundamental idea of atoms underwent profound and fundamental reinterpretation (as we shall see), it is already clear that it raises the question about identity in a new way. If, in fact, the atoms are truly simple bodies, different from the complex bodies they compose, it is possible that atoms are not individuated in the way that large-scale objects are. In fact, Democritus's infinitely varied atoms call out for description and explanation. Anaxagoras thought that a bone is made of tiny bones, a rock of tiny rocks; such an explanation is obviously circular. Aristotle's objection to the strangeness of atoms points to their explanatory power: atomic theory makes phenomena more intelligible, but at the price of introducing unobservable beings in a substratum hidden behind the visible world.

At the very least, any explanation in terms of hidden atoms puts in question the conception of individuality as a sacramental quality inhering in a certain person or object because of the sacred union between that individual and its unique traits. If all material objects have the same atoms, there is no place for "primitive thisness" on the atomic level. The difference between objects calls for a new discussion of form, now understood as the architecture of atomic building-blocks. In turn, this requires understanding exactly how the blocks are put together and how they interact. Turning back to the atoms themselves, this new level of reality raises even more profound philosophical questions.

If atoms can move around and give the appearance of change, there must be void space between the atoms, so that they will have room to move. Here the atomists daringly postulate that the void exists, despite the seeming contradictoriness of such an assertion: how can nonexistence exist? The price of granting the reality of change is to grant nothingness a kind of reality also. After all, if change is real, the thing that exists and is changing is somehow being transformed into something else, into its opposite. The rose is losing its redness, for in a few days it will have it no longer. If that is so, then being and nonbeing must be strangely intermixed. The interplay of atoms and the void accomplishes this mixing.

Abysmal questions follow from this paradoxical claim. The precise being of an atom includes its immutable shape and size, but how can one speak of the "shape" or "size" of the void? Whatever "reality" means, somehow it has to cover both the atom and the void. The objections that Aristotle and others directed to atomic theory pointed to these paradoxes and questions. The difficulties that surround these ideas remained potent even as the atomic theory triumphed, many centuries later. Einstein laid to rest the conception of a material ether filling empty space. Yet soon afterward, quantum theory showed that seemingly empty space could no longer be considered absolutely empty. Like all genuine questions, the paradoxes of the atom and the void have not been decisively answered or refuted; they point to deep issues that must be continually revisited.

In particular, the reality of atoms opens the question of their individuality and distinguishability. Atoms are primitive by definition; the atoms in a rose are simple, featureless constituents. Even at this early stage, the realm of the atom discloses a new level of being, new in kind even more than in sheer smallness. By the same token, atoms raise the question of individuality in a new way and with new intensity. If forms are merely the by-product of the arrangement of atoms, perhaps some common notions should not be taken at face value. Anything not palpably material and observable could be an illusion, compared to the solid reality of atoms. That, at least, is one obvious way of reading Democritus's aphorism that "by convention there is sweet, by convention there is bitter, by convention hot and cold, by convention color: but in reality there are only atoms and the void." This "convention" could swallow many common human certainties, including the conventional gods. If the ship of Theseus did not contain its original atoms, it was a pious fraud.

Both ancient atomists and their critics realized that the strict reduction of the world to atoms and the void challenged conventional teachings about the human soul and the gods. Were all such conceptions merely insubstantial stories, nothing more? That, at least, seemed the implications of atomism to the outraged Athenians who accused Socrates of being one of those godless sophists speculating wildly about the cosmos and teaching impiety to the young.

Even though Socrates denied this accusation, he did not deny that the sophists' explanations would tend to undermine many commonplaces about the soul and the gods. Indeed, the partisans of atomism rejoiced in rejecting what they considered conventional superstition. Their most eloquent advocate was the Roman poet Lucretius, expressing the teachings of his Greek master, Epicurus. Lucretius's poem *On the Nature of Things* glories in showing how atomism liberates men from the fear of death and the terror of the gods. He depicts the horrors of archaic religion, such as the sacrifice of children to appease angry divinities, and contrasts these with the cheerful calm that atomic contemplation can bring. He argues that even the nothingness that follows bodily death is far more comforting than the dreary eternity of Hades, the shadowy realm of boredom that awaited the ancient dead. Though he represents himself as a loyal son of Rome, no great leap separates the rejection of conventional divinity from questioning the divine pretensions of Roman emperors.

Such dangerous assertions doubtless would have inflamed the young and alarmed the conservative. In his prison cell, Socrates rejected offers to escape precisely in order to show the Athenians that the philosopher is a loyal citizen of his city, even after it condemned him to death. He accepts the Athenians' judgments even when they err terribly; he is serene facing imminent death through their injustice. His final conversation shows him persuading his grief-stricken friends that death is not an evil and that immortality awaits the soul. This unforgettable discussion takes place under the aegis of Theseus's ship, for Socrates' execution had been stayed until the return of that ship from its sacred journey, reenacting the voyage of the Minotaur-slayer. As Jacob Klein observed, in this dialogue Socrates appears as another Theseus, seeking to deliver the Athenians from a still greater Minotaur, the fear of death. Yet that very reenactment raises troubling questions. Was the ship then returning from Delos to Athens really the same as the one Theseus had sailed? And were the fourteen Athenians talking with Socrates really the same as the youths and maidens who had accompanied Theseus? In Plato's new philosophical drama, was Socrates really the same as Theseus?

Most troubling of all, even if the soul somehow continues to exist after the death of the body, is it really the same soul? Or is it an equivocal reenactment, like the ship of Theseus? What, then, of Socrates reenacting Theseus, or of us, the readers, who seem to reexperience Socrates' final hours? Everything turns on whether the re-creation really departs from the authentic original. This question is also very much present to Plato as he re-creates the scene in the Athenian prison. In Plato's dialogue, Socrates' disciple Phaedo converses with another devoted follower, Echecrates, who begins their discussion by asking Phaedo whether he had really been there himself or heard it from another. Phaedo was an eyewitness; their conversation is a reenactment of the death-scene. Echecrates lives outside Athens; he lacks detailed information on the death of Socrates, as if the memory of the master were already fading into fragments.

Their discussion seems to bring Socrates back to them, as the ship of Theseus seems to return again, renewed.

In fact, Echecrates had never heard of the ship of Theseus. Phaedo tells him: "The Athenians say that it is the one in which Theseus sailed away to Crete with the seven youths and seven maidens, and saved their lives and his own as well." Phaedo himself was not born an Athenian; he was from Elis, later captured, enslaved, and ransomed. It is hard to gauge his tone of voice as he tells what "the Athenians say." Does he imply doubt in the pious myth, or is he recounting an alien custom faithfully? Phaedo, as a former slave, would know something of the varying customs of men and even more about whether he himself was the same when enslaved or liberated. The question returns to haunt us. Our life, our continuity, our immortality is at stake in the conversation of these characters from an ancient dialogue, long-dead or revived as they may be. A sense of hidden depths surrounds this passage, as if it calls for the famous Delian divers, well known for plunging into the deep waters around the sacred island. Surely a Delian diver is needed to plumb the mysterious identity of the ship of Theseus. No less mysterious is the individuality of each person or each atom.

3
Atoms and Monads

The atomic theory underwent many changes before it came into its own, many centuries after it was introduced by the Greeks. These changes involved new views about the individuality of atoms. In the later ancient world, philosophers like Porphyry cited Aristotle to argue that individuality was really indivisibility. If so, atoms would be individuals because they were indivisible. The Roman Stoics reconsidered the basic question of individuality as the Roman Empire grew ever vaster, dwarfing its innumerable citizens. Cicero thought that "no hair or no grain of sand is in all respects the same as another hair or grain of sand." To be sure, the Stoics held to the old idea of four basic elements. Seneca added that "god has created all the great number of leaves that we behold: each, however, is stamped with its special pattern. All the many animals: none resembles another in size—always some difference! The Creator has set himself the task of making unlike and unequal things that are different." Here, the Stoics' basic principle was that the world is ruled by reason. If there were two identical individuals, there would be no reason for one of them to be *here* and the other *there,* rather than the other way around. They held that each individual has its own uniqueness, its tiny but solitary dignity, though lost in the vastness of the Empire. As G. W. F. Hegel wrote, "Stoicism could only appear on the scene in a time of universal fear and bondage, but also a time of universal culture which had raised itself to the level of thought." The belief in individuality would comfort those struggling to bear their lot; utter anonymity would crush them.

Here a certain religious element enters into these questions, reflecting a wish for the deity to single out and protect increasingly faceless and desperate multitudes. For instance, the Christian philosopher Boethius argued that individuality is a kind of uniqueness. In Greek times, atomism tended to be suspected of impiety. Later, the atomic theory proved no less suspect in certain Christian eyes. All the more surprising, then, that an Islamic school welcomed the atomic hypothesis already in the tenth century after Christ. The Mu'tazilī *mutakallimūn,* the philosophers of *kalām* (meaning speculative theology), argued that not only matter but space itself was atomic. What is more, they held that "all these [atoms] are alike and similar to one another, there being no difference between them in any respect whatever," as Moses Maimonides phrased it, reporting their teachings in his *Guide of the Perplexed.* This is especially unusual, given that most Arabic philosophers revered Aristotle as *al-falusūf* ("the philosopher" *par excellence*) and echoed his anti-atomism.

The advocates of *kalām* took this radical position in order to articulate an equally radical view of God. To emphasize divine power, advocates of *kalām* argued that the world could not be ruled by simple mechanical causality, which would leave God with no essential role in the running of the cosmos. Instead, they held that God constantly interferes in the operation of the atoms so as to maintain the order that we observe, creating and annihilating them as He pleases. As befits the mere tools of the divine ruler, the atoms are unindividuated; their utter lack of individuality exalts His creative power.

Kalām was unusual in the Muslim world in its attitude toward atomism. Many theologians, whether Muslim or Christian, thought that it smacked of atheism. It particularly disturbed Christian believers in the transubstantiation of bread and wine. How could the same atoms be now wine, now blood, without any apparent change? Atomic theory threatened the reality of the most revered Christian sacrament, which rests on an invisible transfiguration of identity. Hence, some Church Fathers called the ancient atomists "swine." In 1415, the Council of Constance condemned John Wyclif for the heresy of "Epicureanism" (as atomic theory was called). Later, Giordano Bruno embraced atomism in his radical vision of an infinite cosmos that followed Copernicus rather than Aristotle. Contrariwise, one of the merits of Aristotle's philosophy was its rejection of the troublesome atom. Among Muslims, the renowned al-Ghazali turned to Aristotle as he mounted his devastating attack on the *kalām* in the eleventh century, after which that school withered away.

Yet atomic theory did not succumb to dogmatic rejection. In the West, controversy about the status of the Eucharist during the Protestant Reformation may have helped dispel the cloud over Epicureanism. Though Francis Bacon was critical of atomism, he nevertheless thought it came closer to the secrets of matter than any other ancient philosophy. In France, a Roman Catholic priest, Pierre Gassendi, was instrumental in reviving Epicurus's philosophy during the 1640s and even making it fashionable. To do so, he mitigated aspects of the theory that were theologically obnoxious, making the atoms creations of God, not eternal, and emphasizing that they serve divine purpose without limiting divine power.

Gassendi was a central figure in the new science, a tireless correspondent whose influence drew wide attention to atomism. He compared the different sorts of atoms to letters of the alphabet, implying that the atoms of each kind are as identical as letters. Here the uniform atoms of the *kalām* find a kind of reconciliation with the endless diversity of Democritus's atoms. Then Gassendi extended the alphabetical metaphor: "From the atoms the smallest molecules are joined together first, and then successively somewhat bigger ones, still bigger ones, the finest and the coarsest bodies, and finally the biggest bodies." Here the concept of *molecule* emerges, showing how a limited number of atomic species might lead to a much larger variety of molecular structures underlying visible matter in its myriad forms.

Among Gassendi's correspondents, René Descartes did not adopt a simple atomism, unconvinced of the existence of a void between atoms. However, Descartes developed an account of the smallest particles of matter that casts an interesting light on atomic theory. In the beginning, Descartes argued, space was filled with matter composed of particles of uniform size, all moving and packed together with no void space between them. This required that the motion of primordial matter was circular, turning endlessly as if it were a cosmic mill-wheel (figure 3.1). As a result of the constant contact and grinding between adjacent particles, this cosmic mill-wheel would slowly grind them down, yielding different sizes of particles, coarse, medium, and fine, like the grades of flour from a mill. In the process of painting this picture, Descartes also raised the possibility that atoms might not be impenetrable but instead subject to wear.

By the time of Isaac Newton, atoms were a familiar element of the new natural philosophy. Though Newton's physics de-

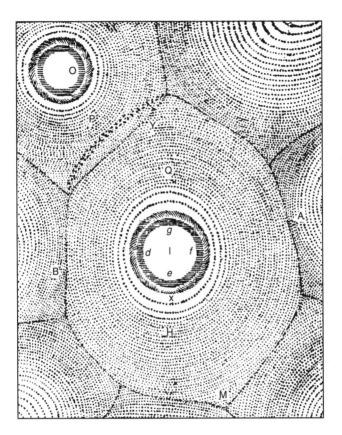

Figure 3.1
René Descartes's cosmic mill; matter starting at A is ground down as
it moves around the central star, I, through the cosmic vortex AMBY.
O is a neighboring star. (*Principles de Philosophie;* Paris, 1647.)

parted greatly from the ancients, his conception of the atom
was still close to theirs.

All these things being consider'd, it seems probable to me, that God
in the Beginning formed Matter in solid, massy, hard, impenetrable,
movable Particles . . . and that these primitive Particles being Solids,

are incomparably harder than any porous Bodies compounded of them; even so very hard, as never to wear or break in pieces; no ordinary Power being able to divide what God himself made one in the first Creation.

Newton enshrined this perfect hardness in his "Rules for Philosophizing." Unlike Descartes, Newton held that matter could not be ground ever finer. Newton insisted that inertia was the essence of matter, not merely volume or extension, thus placing the concept of mass into the foundations of the new science along with hard, massy atoms. Robert Boyle also thought that a body was just "a distinct portion of matter which a number of [atoms] make up." Deeply influenced by Gassendi, Boyle set out to find experimental verifications of the atomic hypothesis in the laws of gases.

Yet Boyle also wrote that "it is no such easy way as at first it seems, to determine what is absolutely necessary and but sufficient to make a portion of matter, considered at different times or places, to be fit to be reputed the same body." This is the precise question of identity, the problem of how atoms retain their individuality in all different times and places, if they are perfectly hard. John Locke, Boyle's younger contemporary, thought that the identity of an atom depended on its location in space and time, for "what complexion of accidents besides those of place & perhaps time can distinguish two atoms perfectly solid & round & of the same diameter?" If atoms lack any distinguishing qualities, they still have definite locations at every instant. That, in turn, would guarantee their continuing identity over time, for "'tis evident, that, considered in any instant of its Existence, [an atom] is, in that instant, the same with it self. For being, at that instant, what it is, and nothing else, it is the same, and so must continue, as long as its Existence is continued: for so long it will be the same, and no other."

The atomic theory shifted the question of identity from bodies to the atoms constituting them. Yet what keeps each atom's individuality from changing over time, perhaps by some matter rubbing off as in Descartes' cosmic mill? This question will return, though in the seventeenth century the presumption of atomic indivisibility remained unquestioned. That being presumed, Locke saw no problem in the continuing identity of atoms but had more difficulty with human identity. Thinking about the ship of Theseus, one could ask whether the physical change in the human body over time, including the gradual replacement of tissues, undermines the identity of the human being. In order to avoid such problems, Locke asserted that a man is not just his atoms, but rather his consciousness and memory, which connect childhood with age and connect a person asleep with the same one awake.

Yet even in the case of the ship there are some strange possibilities. Thomas Hobbes had noted that "if some man had kept the old planks as they were taken out, and by putting them afterwards together in the same order, had again made a ship of them, this, without doubt, had also been the same numerical ship with that which was at the beginning; and so there would have been two ships numerically the same, which is absurd." Which of the two ships was the real ship, the one with the original planks or the one with the exact replacements? The replacement (or the duplication) of the ship strikes at the "thisness" of the original vessel.

Locke imagines an even more disturbing problem when this is applied to a human being. "If the same consciousness (which, as has been shown, is quite a different thing from the same numerical figure or motion in body) can be transferred from one thinking substance to another, it will be possible that two thinking substances may make but one person." Two separate beings might actually be one person, if they shared memories and consciousness. Locke imagines the soul of Castor transplanted into the body of Pollux: "the Bodies of two men with only one Soul between them, which we will suppose to sleep and wake by turns; and the Soul still thinking in the waking Man, whereof the sleeping man is never conscious, has never the least Perception." Thus a single soul might animate two bodies alternately, one awake while the other sleeps. Bizarre as it may seem, this strange possibility cannot be discounted, as the next chapter will consider.

Locke also considers the significance of a breakdown in memory, as in total amnesia. His answer is radical: "If it be possible for the same Man to have distinct incommunicable consciousness at different times, *it is past doubt the same Man would at different times make different Persons.*" For instance, consider a general who remembers his memories when he was a young lieutenant, but not the early childhood memories that he as a lieutenant still recalled. Locke would consider the child and the general to be different persons, though the lieutenant shared memories with both of them and all three were the same man. This is consistent with Locke's insistence that personhood is a moral state whose rights and responsibilities are meaningless if the person cannot remember his or her acts.

Nevertheless, Locke's treatment of identity surprised its readers. Besides the apparent paradox of the same man ceasing to be the same person, there were even more volatile theological issues. The resurrection of the dead, body and soul, is a tenet of Christianity, but in what sense were those who had risen the same as they had been in life? Some held that the risen body had to be identical with the body before death, atom for atom. Though a staggering task to contemplate, the infinite God "can with as great facility re-unite these dispersed *atomes,* as he could at first create them," as Alexander Ross put it. Others held that the body could be the same without every particle being identical, referring to rabbinical speculation about a *luz* or "resurrection-bone" in the body that was not destroyed by death and whose unbroken continuity maintained identity. As Boyle put it, "there is no determinate bulk or size that is neces-sary to make a human body pass for the *same.*" He moved toward a third position, that only the continuance of the soul is required for resurrection. As Kenelm Digby argued, "If *God* should joyne the *Soule* of a lately dead man . . . unto a *Body* made of earth taken from some mountaine in *America*; it were most true and certain that the body he should then live by, were the same Identicall body he lived with before his *Death* & late *Resurrection.*" Here the continuity of the body is guaranteed by the sameness of the soul. Locke's position seemed close to this, for he identified personhood with continuity of memory and soul.

But this position was not acceptable to Edward Stillingfleet, bishop of Worcester, who attacked Locke's position as incon-sistent with the resurrection of the body and hence with the Christian faith. Besides giving theological objections, the bishop argued that location in space and time cannot suffice to distinguish the individuality of bodies. Though our knowledge may depend on location to distinguish things, we should not confuse our ability to locate things with the very essence of those things. Distinguishability has to do with our knowledge and discernment, whereas identity concerns the "thisness" of each thing in itself, beyond our ability to discern it. Homer al-ready showed that, though Odysseus might seem indistinguish-able from an ordinary beggar, underneath he remained the same man who built the Trojan horse, whose uniqueness was intact under his disguises.

This argument continued to resound, especially its plausible assumption that each body had to have some intrinsic, unique "thisness" as the touchstone of its very existence, on which rest all distinctions of place and time. In turn, such a deep-seated "thisness" implied that each individual had to be utterly different from every other. However, this argument may well not extend from persons to things. Though human beings seem necessarily unique, it is less clear whether there might not be two identical blades of grass or two identical atoms.

The question of finding two identical individuals was the subject of a strange bet made on a summer day about 1695. In the gardens of Princess Sophie in Herrenhausen, an argument was underway between the renowned philosopher, diplomat, and mathematician Gottfried Wilhelm Leibniz and an "ingenious gentleman." Against Locke, Leibniz had maintained that there is no such thing as two identical individuals, each indiscernible from the other. The gentleman asserted that there were such perfect twins, and the princess defied him to find two leaves in the garden that were exactly alike. The gentleman searched for a long time, going all over the garden, but without finding the two matching leaves he swore must be there. Leibniz felt vindicated and told the story many times in his writings.

Although the ingenious gentleman thought it strange, later generations of children were commonly taught that no two snowflakes are ever exactly the same. After the debacle in the garden at Herrenhausen, many curious eyes have watched for that mysterious prodigy, a perfectly matched pair of snowflakes, alike in every point. Experience seemed to bear out Leibniz's contention, which he named the *Principle of the Identity of Indiscernibles:* "no two substances are completely similar, or differ only in number." This means that if we *thought* we saw two seemingly identical snowflakes, we must have looked twice at the same individual.

Quite apart from experience, though, Leibniz also had a logical proof of this principle that goes back to the Stoics' arguments: If there were two truly identical objects A and B, there would be no reason for God to put A *here* and B *there,* and so God (always doing things for a sufficient reason) could never have created such beings. Moreover, Leibniz noted the intimate connection between his principle and the axiom that no two bodies can occupy the same place at the same time; if so, the identity of a body corresponds to its impenetrability. Also, Leibniz held that if every being is unique, the world is continuous and unbroken; nature does not make jumps but always moves smoothly. These assertions of impenetrability and continuity are crucial consequences of individual identity that will later return in a very different key.

For people no less than for atoms, the question of identity has profound effects on the smallest and largest scales. Leibniz's Principle means that every single substance in nature is an individual, set off from all others by its incommunicable individuality, so that it is in some respect unique. Although Leibniz had been thinking about leaves or other objects of normal size, he extended it further: "Two drops of water, or milk, viewed with a microscope, will appear distinguishable from each other." If this principle held on the atomic scale, no two atoms would be exactly the same. To be sure, Leibniz rejected atoms and noted that his principle would contradict the notion of unalterable, indistinguishable atoms: "In fact, however, every body is changeable and indeed is actually changing all the time, so that it differs in itself from every other. . . . That is why the notion of atoms is chimerical and arises only from men's incomplete conceptions."

Leibniz's Principle seems commonsensical; it appeals to our everyday experience even of manufactured objects, which always have some slight difference between them, though cast out of the same mold. The force of his argument is even stronger if one considers only natural or hand-made objects, which have an obvious individuality that is minimized in modern manufacture; the subtle but telling differences between identical twins comes to mind. Still, one hesitates, wondering whether one might find, on some exceptional occasion, two identical snowflakes, if not two identical people. Somehow, the smaller the object, the more it seems to us that it might lack individuality, if only because our ability to discern and distinguish would fail us. If truly indistinguishable objects really exist, they would be found in a realm foreign to normal, everyday experience, perhaps most likely in the realm of the very small.

Individuality indicates not only what is familiar, but what seems human. For most people, though there are limits beyond which individuality verges into eccentricity or idiosyncrasy, to be "faceless" is a strong reproach. It indicates an undifferentiated member of a crowd, someone who is less than human in our eyes, perhaps through unfamiliarity, though the implication is that the faceless person never earned a face, as one who has "lost face" has been dishonored. Certainly most people would regard themselves as uniquely identifiable, not replaceable; the ceremony of marriage, for instance, requires the identification of the names and persons being wed, as a matter of course. Indeed, having a name is for humans the sign of individuality; pets and even cars or machines acquire names as they become uniquely identifiable for us; the giving of names was Adam's first task.

When Leibniz argued that every being is completely individual, he was much influenced by the example of living beings; furthermore, in his view, every being *is* living, in the sense of having a unique career that seems to reflect all the other bodies that impinge on it. Rather than considering atoms as the elementary constituents of the world, Leibniz conceived of a world of radical individuals, which he called *monads,* each completely unique and yet all coordinated together in what he called "preestablished harmony," a mysterious choreography that connects the career of each separate monad to all the others without any of them actually reaching out and influencing its fellows. These monads are not defined by size in the same way atoms are; Leibniz envisioned an infinite hierarchy of monads within monads. Here he was deeply influenced by looking through early microscopes and seeing the complex world of minute creatures inside a drop of water. To Leibniz, each creature in the microworld seemed to have a full and complex life, no less than the creatures of the macroworld. He considered this preestablished harmony to be the way that God coordinated the myriad individuals into a whole without performing obvious acts of divine intervention. This daring conception necessitated replacing faceless atoms with individual monads. It represents his extreme adherence to a world of individuals, whose very life is inseparable from their individuality.

Turning from microscopic monads to humans, it is clear that Leibniz's vision amounts to a striking assertion of the importance of individual human character. Here, too, Leibniz invoked divine coordination, arguing that a good God must have arranged events in the best way possible, meaning not mindless optimism but that the world was as good as it could be, in the sense of making the best out what is really possible. How could that be, if individuality could be effaced? To this cauldron of deep questions, David Hume added a final explosive ingredient, coming from the very different context of Scottish philosophy with its strong empirical bent. Stepping back from Leibniz's sweeping assumption of universal harmony, Hume argued that the mind is in reality "nothing but a heap or collection of different perceptions," which we endow with coherence and persistent identity only by mere habit. Our imagination makes us ascribe successive sensations to what we presume to be an identical object, but whose true individuality may be utterly incoherent. Thus the identity of the ship of Theseus is a quality of our imagination, nothing more. Our individual identity also dissolves into a theater of successive perceptions, given unity only by our imagination. The conviction that each of us is some one, unique, enduring individual is a flight of fancy over an abyss of uncertainty.

Hume's arguments have continued to resound in the minds of thoughtful people, and his own response was memorable. He depicts himself as struck with melancholy, a castaway living on a barren rock surrounded by a boundless ocean of uncertainty on which he faces imminent shipwreck. He felt himself to be a kind of monster who had infected his fellow thinkers with crushing doubts.

I dine, I play a game of back-gammon, I converse, and am merry with my friends; and when after three or four hour's amusement, I wou'd return to these speculations, they appear so cold, and strain'd, and ridiculous, that I cannot find in my heart to enter into them any farther. Here then I find myself absolutely and necessarily determin'd to live, and talk, and act like other people in the common affairs of life. . . . I am ready to throw all my books and papers into the fire, and resolve never more to renounce the pleasures of life for the sake of reasoning and philosophy. . . . If I must be a fool, as all those who reason or believe any thing *certainly* are, my follies shall at least be natural and agreeable.

If all we have is habit and imagination, perhaps reimmersion in common human reality might save us from these dark thoughts. However, even the realm of the senses may not be the safe haven it seems.

4
Secret Sharers

Up to this point, there have been two alternatives: either individuality is a sacrament that cannot be shared or it is a commodity that can be exchanged, akin to coins of the same denomination. These alternatives seem to have no common ground. However, a strange story from sixteenth-century France opens up a new perspective on the nature of identity. Leibniz already knew of this case and mentioned "the false Martin Guerre" in the course of his discussion of individuality. It was, indeed, a celebrated legal case that has been retold over the centuries, most movingly in Janet Lewis's short novel *The Wife of Martin Guerre.*

Briefly, in a remote village in southern France two villagers, Martin Guerre and Bertrande de Rols, were married in 1539, when each was 9 or 10 years of age; such early marriage was not uncommon at the time. They both came from well-established families in the region and had a son together. About ten years later, Martin disappeared after angering his father (he had stolen some wheat from him). For eight years, he did not return; during that time his father died. Then Martin reappeared, to the great rejoicing of his family. They all recognized him, though he seemed changed; Lewis depicts him as much gentler than when he left, not as laconic and severe. He and Bertrande were about to have another child when she finally realized that he was not really Martin Guerre. The historical records point to his uncle Pierre as the moving force that began legal proceedings against Martin for fraud and misrepresentation; Lewis places the burden on Bertrande herself, gradually convinced that this man was not really her husband, even though he was in many respects better than the old Martin.

A lengthy trial ensued; "Martin" was declared guilty on the testimony of his wife, even though many other witnesses (including his sisters) stood by his story. He was condemned to death, to Bertrande's horror, for she had not asked for the extreme penalty; the case was appealed and the pressure mounted on her to relent, to spare what seemed a decent man and the head of the family. She remained steadfast; at the end of the appeal the court was perplexed, with many witnesses on both sides (some agreeing with Bertrande and identifying the defendant as the rogue Arnauld du Tilh, others persisting in identifying him as Martin). The appellate court was on the verge of declaring him to be the real Martin Guerre. At the very last moment, a one-legged stranger appeared, claiming to be the real Martin. He was confronted with all the various witnesses, one by one, who recognized him with varying degrees of shock and chagrin. Finally, Bertrande was brought in. She, too, recognized him, begging his forgiveness for her initial mistake, pleading both her own desire to see him return as well as the universal assent of his kindred to the false Martin. The newly arrived Martin was cold and implacable, true to his old severity; he would not forgive his wife for the dishonor she had brought to his house (figure 4.1). At that point, the extant legal records end, save to record the later confession and execution of the imposter Arnauld du Tilh. Lewis depicts Bertrande's departure from the courtroom:

Figure 4.1
An early illustration of the case of Martin Guerre, showing the arrival of the peg-legged stranger. (Jacob Cats, *'S werelts begin, midden, eynde;* Amsterdam, 1663.)

The court did not detain her, and the crowd, in some awe, drew aside to let her pass without interruption. Bertrande did not see the crowd. Leaving the love which she had rejected because it was forbidden, and the love which had rejected her, she walked through a great emptiness to the door, knowing that the return of Martin Guerre would in no measure compensate for the death of Arnauld, but knowing herself at last free, in her bitter, solitary justice, of both passions and of both men.

Jean de Coras, the eminent jurist who was among the appellate judges and who recorded this incredible case, writes at the end of his account *A RAISON CEDE,* "yield to reason." Perhaps he meant that, though this case is more bizarre than fiction, in some way one's reason should attempt to encompass it. The domain of reason may be large enough to include such events that seem to defy common sense. Or perhaps he is recording his bafflement.

In many ways, the story of Martin Guerre twists the story of the homecoming of Odysseus. In place of the real hero returning disguised as a beggar, "Martin" is a scheming fake, a gifted mimic possessing a phenomenal resemblance to the real Martin, who has somehow gleaned crucial tidbits of information that will render him irresistible. Did the real and false Martins collude? In what other way could the imposter divine or discover those intimate secrets? It is hard to resist the impression that both men are guilty of a bizarre and predatory conspiracy, though both denied it. If the false Martin acted alone, how did he manage his masquerade? What tricks could enable him to counterfeit the other's very self?

In the end, the first court was justified in thinking that a man's wife would know him from all others, and the events bore them out. Betrande's situation is far more complex than Martin's, and she is rightly the center of Lewis's version of the story, in which she is not implicated in the plot. Alternatively, a modern scholar has argued that Bertrande secretly recognized the "false" Martin, preferred him to the "true" one, and conspired with him to keep the secret. Yet would she then turn on him and prosecute him for giving her the very things she wanted, their self-created marriage defying law and custom? Could Pierre have had such power to compel her to act against her own interest and choice? Individuality may not be so patent. Bertrande could well have been caught up in "Martin's" convincing performance, which was further confirmed by their friends and kinsmen (too numerous to be part of any conspiracy). Ironically, Lewis indicates that the new Martin was too nice to be real; the persuasiveness that first helped him finally betrayed him. The eloquence he needed to charm and reassure was deeply out of character with his callous, terse double.

In any case, Arnauld was able to step across the border of individuality. It is as if a false Odysseus were to have deceived a trusting Penelope to recognize him as the real man. Indeed, Penelope seemed to fear just such a possibility; she knew that her own overwhelming desire to see her husband return might deceive her. Even if everyone around her were deceived, she must not be. Penelope knew not to trust even the physical signs, the scar that Odysseus bore from a youthful hunting accident and that revealed him to his old nursemaid. She trusted in their intimate secrets; she also trusted that Odysseus would not reveal them to anyone else. The story of Martin Guerre casts a dark light on this ancient recognition. The imposter had the right scars (as if in ironic recollection of the case of Odysseus) and even knew the intimate secrets. Penelope's fear, and Bertrande's subsequent experience, confirms that there is a danger of confusing or merging individualities that might deceive the most faithful wife. This goes considerably beyond the illusion projected by a skillful actor, who persuades his audience that he *is* his character, although already that requires a certain suspension of the ordinary limits of identity. More than the most committed actor, Arnauld had to "live" his part so deeply that there was no question of impersonation. Stranger still, this deception happened not on stage but in "real life."

This story opens up new and disturbing possibilities of merged individuality. At first, the false Martin was more "real" than the one-legged man he replaced and in the ways that mattered most to his wife and family. There are important senses in which the "real" Martin, so implacable and cruel, did not deserve to return. Many divorced persons would hold that their former spouse had, over time, ceased to be the person that they had originally married and had become for them unbearably false. There is a certain strange fluidity to human identity, even when there is no question of imposture. At the end of the *Odyssey*, one anticipates a prolonged period of reacquaintance so that Penelope and Telemachus can come to know the man they have not seen for twenty years (and whom Telemachus never knew), for Odysseus must have been deeply affected by his experiences during that time. Their mutual recognition does not exclude a certain strangeness. This applies even more to Bertrande, now reunited (at her own behest, through great upheaval) with the ultimate stranger, her rightful husband. Odysseus's family seems reconciled to his strangeness; at the end, it is hard to imagine Bertrande being reconciled to Martin new or old, in her bitter and solitary freedom.

In some ways, one could take the story of Martin Guerre to show that the indelible "thisness" of individuality will eventually emerge, however unhappy may be its result. Despite the confusion, the climactic confrontation decisively unmasked the imposter and affirmed the true Martin. Thus, one might take the confusion as temporary and, in the end, not significant. However bitter her victory, Bertrande finally exorcised the demon that haunted her, despite the feelings of those around her and her own initial reaction. However, the fortuitous appearance of the real Martin does not dissolve the troubling doubt: what if he had not come back, then or ever? How was she to be certain that she was right to demand the death of her once-beloved spouse? How was the court to decide on a matter that ultimately rested in her inmost soul or (worse still) was coerced from her by uncle Pierre, who wanted to see Martin ousted so that he could become the head of the family? Jean de Coras's account shows how much these things perplexed and troubled the judge; after all, the "imposter" Martin long denied the one-legged man's assertions and in fact seemed to have a better command of the couples' intimate secrets than the newcomer.

Coras thanks God for resolving the unparalleled perplexity of the court, "showing that He always wishes to help justice and that such a prodigious matter should not remain hidden and unpunished." He knew that this divine intervention did not dissolve the deeper issues. Only twelve years after the Guerre trial, Coras's own identity as an eminent jurist and Huguenot was destroyed when a mob hanged him in his scarlet judge's robes during the aftermath of the St. Bartholomew's Day massacre. Many of his contemporaries continued to meditate on this disturbing case; the jurist Estienne Pasquier asked whether Martin Guerre, who was so harsh toward his wife, did not deserve as severe a punishment as Arnauld, since his absence had caused the wrongdoing. Michel de Montaigne thought that, despite the end of the trial, the evidence still remained inconclusive; had not sixty of the witnesses been unable to tell between the two Martins? How could Coras have been so sure as to condemn du Tilh to death, rather than to imprisonment or the galleys? Montaigne's doubts were not removed by Arnauld's confession, for "such persons have sometimes been known to accuse themselves of having killed people who were found to be alive and healthy." Perhaps he confessed out of a kind of gallantry, to ease the lot of Bertrande, whom (Lewis indicates) he had come to love. In any case, Montaigne takes the story to warn against human presumption; he would have preferred the verdict "The court understands nothing of the matter," recalling the ancient tribunals who, in perplexity, would sometimes adjourn a difficult case for a hundred years, awaiting full clarity.

Though doubtless Coras acted within the judicial norms of his time and in the face of pressing need to resolve the practical question, Montaigne indicates the grounds that have kept this case alive in the imaginations of succeeding generations. Even the possibility of such questions about individuality opens the door to a strange realm of uncertainty. In several plays, Shakespeare uses stories of mistaken identity to unlock comic possibilities. Two pairs of identical twins, masters and servants, cause farcical confusion in *The Comedy of Errors.* The possibility of confusion of identity excites laughter, but touches on a deeper quest. As one of the twins says,

I to the world am like a drop of water
That in the ocean seeks another drop,
Who, falling there to find his fellow forth,
(Unseen, inquisitive) confounds himself.

The resonance of mingled identity also joins the brother and sister Sebastian and Viola in *Twelfth Night.* A darker mirroring connects the inseparable friends Valentine and Proteus in *The Two Gentlemen of Verona,* as Proteus turns toward betrayal of Valentine, his other self.

Later stories of doubles amplify these disturbing possibilities. The encounter between a *Doppelgänger* (an independent yet identical version of a person) and his alter ego is an important element of Romantic literature, uncanny and evocative; its touchstone is Charles Baudelaire's poem "To the Reader," which concludes with the poet addressing his "hypocritical reader" as "my double, my brother" *(mon semblable, mon frère).* The boundary between author and reader may be transgressed. Beyond literary artifice, there is a long-standing folk belief that each person has somewhere in the world a double, whom perhaps they might hear spoken of (by a friend who has met both) or even catch sight of. However, actually to meet and know one's double is said to be dangerous or even disastrous, as if the confrontation of two identical selves were so contrary to nature that it leads to horrible consequences. It is said that one or the other of the pair must die as a result of their ill-fated encounter, as if to restore the balance of nature by eliminating from the world the monstrosity of doubled identity. Perhaps this also represents a kind of primal incest, the commingling of separate identities against the prevalent order of nature.

Fyodor Dostoyevsky's novella *The Double* consummates this dark vision. Golyadkin, a nervous and timid civil-service clerk, follows a stranger and eventually recognizes him as his double, who goes into Golyadkin's own apartment and appears right next to him at the office. At first, he thinks it a dream, or perhaps a monstrous joke. "He even began to doubt his own existence. . . . His misery was poignant and overwhelming. At times he lost all power of thought and memory." His coworkers do not seem to notice, though one of them gradually admits that the resemblance is amazing and speaks of Siamese twins. Golyadkin Senior (as Dostoyevsky calls the original man) gradually begins to converse with the newly appeared Golyadkin Junior, who seems to be even more mousy and abject than Senior. Senior offers Junior dinner and drinks; they become relaxed and sentimental and Senior invites Junior to stay on as a guest. The next morning they have mysteriously exchanged identities; the servant thinks that Junior is his master and Senior is the guest. As the day unfolds, Junior slyly usurps Senior's place at work and receives praise for what Senior has done, while Senior is suddenly the object of suspicion and anger from his superiors. Gradually, Senior decides he cannot stand being treated like a "rag." In a series of nightmarish scenes, Junior sticks Senior with a restaurant bill and leaves him to delirious dreams that "a terrible multitude of duplicates had sprung into being, so that the whole town was obstructed at last by duplicate Golyadkins." In the end, Senior loses his beloved to Junior, who is a master of perfidious manipulation and toadying. Junior gives Senior the kiss of Judas as a cruel doctor comes to take Senior away to the madhouse.

During this disorienting crescendo, one never knows whether Golyadkin is simply subject to a growing insanity or whether he is the victim of his baleful, evil double. Since psychosis often involves dissolution of identity, the two may well be indistinguishable. These stories of doubled or merged identities ask whether, when distinguishability is utterly lost, individuality might not also disappear. Passing the borders of identity leads to a strange and disquieting realm that may contain both truth and madness.

Accordingly, stories of doubles are often horrific, exciting the *frisson* of the uncanny by evoking a primal terror. Yet there is also the possibility of rapport and revelation, as in Joseph Conrad's novella *The Secret Sharer*. In that story, a young captain encounters his double in the form of an escaped prisoner, a first mate escaping a charge of murder incurred as he acted to save his old ship during a storm. Mysteriously, the captain trusts and protects the naked stranger, sharing clothes and secrets.

He appealed to me as if our experiences had been as identical as our clothes. . . . He was not a bit like me, really; yet, as we stood leaning over my bed-place, whispering side by side, with our dark heads together and our backs to the door, anybody bold enough to open it stealthily would have been treated to the uncanny sight of a double captain busy talking in whispers with his other self.

The danger of discovery and capture takes the captain "as near insanity as any man who has not actually gone over the border," saved only by a sympathetic gesture from his "second self."

They successfully conspire to arrange the double's escape. The captain receives the ultimate accolade from his alter ego:

"As long as I know that you understand," he whispered. "But of course you do. It's a great satisfaction to have got somebody to understand. You seem to have been there on purpose." And in the same whisper, as if we two whenever we talked had to say things to each other which were not fit for the world to hear, he added, "It's very wonderful."

The escape requires the captain to put his ship at great risk. Glimpsing the hat he had generously given floating on the sea, the captain realizes that "it was saving the ship, by serving me for a mark to help out the ignorance of my own strangeness." In that moment, the captain sees his second self "a free man, a swimmer striking out for a new destiny." Through sympathy and mutual identity, both men enter a new realm.

5

Distinguishability and Paradox

Though Leibniz was sufficiently struck by the story of Martin Guerre to mention it, he treated it as a closed case. By the end of the affair, the uniqueness of the person finally stood affirmed, despite the confusion over the imposter. In his philosophy, Leibniz thought that he had settled the question of individuality. Every being, up and down an infinite ladder of size, was a unique monad—the name means a solitaire—with a unique history. This may very well be the immediate source of the commonplace view that each snowflake is unique. But if it is so commonsensical, why did it seem so surprising to the ingenious gentleman in the garden? Why did he have to check for himself to see if all the leaves really were different? Indeed, some children will still go out looking for those perfectly matched snowflakes. The gentleman was, in fact, no *naïf*; he was Carl August von Alvensleben, a well-educated diplomat, sophisticated and worldly.

His incredulity reflects an older, more variegated sense of individuality that is grounded in ancient and medieval philosophy. Though Aristotle considered the individual to be the basis of all classification, he also thought that membership in a species provides a higher kind of identity than isolated selfhood, for no being in nature really exists apart from others of its kind.

It is intelligible only as a member of its species and, more generally, of its genus. Here Aristotle refers to the hierarchical classification of living beings, whose individuality must be located by placing them in the correct kingdom, family, genus, and species. Aristotle thought that everything in the cosmos is drawn by love to a single prime mover, which exists beyond nature and is utterly impassive, unmoved, and unique. No being in nature could have the kind of uniqueness that is the hallmark of the prime mover, which is the source and goal of all motion in the earthly and heavenly realms. This impressive vision of divinity was much admired by Aristotle's Christian readers, especially St. Thomas Aquinas, who sought to wed Aristotle to Christian teaching.

However, Aquinas made some crucial innovations in his treatment of identity. His God has a very different relation to the cosmos than Aristotle's unmoved mover, which is not a creator since Aristotle considered the cosmos to be eternal, not created and limited in time. As creator, the biblical God shaped creatures and gave some of them personhood, an essential quality of His own essence. Aquinas argues that personhood is a special kind of individuality that can be found in rational beings. Furthermore, there are three persons in God, all equal, and yet each having an incommunicable quality that sets it off as a different genus from the others: the Son is distinct from the Father, as is the Holy Spirit.

This incommunicable quality is the mark of all true persons (Aquinas notes), and it raises the question of whether persons differ from each other so deeply that each person is a separate species unto him- or herself, or whether all persons might form a species in the way that all ponderosa pines form a species. This question turns on deep issues of individuality and will prove to be crucial. There is a disturbing issue here that tends to be overlooked: in earlier times, to be a person was a high dignity. "Persona" also denoted the mask worn by a character in a tragic drama.

It is disturbing to think that, in Greek or Roman times, most of us would not have been considered persons, not just in this exalted, tragic sense but even in the barest legal meaning. In Roman law, only the father of a family, the *paterfamilias,* was a person in the eyes of the law; all the members of his household, including his children and slaves, fell under his authority, which gave him absolute right over their lives, including the right to kill them at will. Not to mention the subjection of his slaves or wife or daughters, even his sons did not become persons while the father lived, no matter how old they were, unless their father specifically gave them their personhood. In order to give public form to this ceremony of manumission, early Roman jurists evolved a kind of legal pantomime whose gestures could be read by the illiterate. The father ceremonially slapped his son with a gentle blow on the cheek, turning him around to face the court as a person. This legal theater of a violent encounter between father and son reminds one of the tragic drama that seemed to the Greeks the crucible of personhood. Following this confrontation, the son metaphorically puts on the *persona* by which the court recognizes him as a being of sufficient independent power that he may be heard in legal deliberations.

From this perspective, personhood regains its original, weighty meaning. As Aquinas confronted difficult questions about the nature of angels, he decided that they must share in the superlative individuality of God for they, too, are divine spirits. Thus the kind of individuality Aristotle reserved solely for his prime mover Aquinas extended to the angels. Each angel is a person, whose essence is incommunicable and unique, and each is a species unto itself, not to be confused with any other angel. This conception of individuality as akin to incommunicable personhood represents an important shift away from Porphyry's concept of individuality as indivisibility and a further accentuation of Boethius's concept of individuality as uniqueness. This thirteenth-century development saw the incommunicable personhood of God as the source and model for individuality, at least of angels. In contrast, all ponderosas share a single species, regardless of their different shapes or sizes; they do not have as much individuality as angels. Human beings lie somewhere in between; as animals, they belong to a single species, *Homo sapiens,* but as persons, each is unique.

This, I think, is what the ingenious gentleman may have had in mind when he was startled by Leibniz's claim that each leaf is distinguishable from every other. Leibniz takes the little differences between leaves to be all-important, whereas Aquinas and Aristotle would consider them insignificant compared to their shared identity given by their parent trees. For Aquinas, angels and immortal souls were unique and distinguishable because of their divine provenance. The gentleman may well have found it odd that each leaf could also be unique, since it lacked the exalted dignity of those unique spirits. To be sure, Leibniz certainly did not mean that the leaf had personhood. Yet how can a leaf gain a degree of individuality seemingly appropriate only to persons, something denied to many human beings under Roman law? And why does the individuality of a leaf stem from physical, rather than invisible, distinctions?

For Leibniz, external differences are the manifestations of individuality. However, if I pick up a leaf and put it somewhere else, that does not change its individuality, which is independent of location in space or time. Similarly, I might change the leaf's shape or paint it a different color without altering its individuality or affecting its identity. Moved by these considerations, Leibniz argues that, though not a person, each leaf has an individuality that sets it apart from all other leaves, just as each angel is set apart from all others. In his view, this realization extends to all beings, for he would have been deeply troubled if there were one kind of individuality for some beings, and a quite different kind for others. He rejected a sharp break between rational creatures having individuality and the rest of the world lacking it. For Leibniz, all nature is of a piece, a smooth, seamless whole in which there are no breaks or discontinuities. In his vision of the monads, there is no essential difference between the microscopic world and the macroscopic, or between animate and inanimate. All monads have the complex individuality of living things and they all mirror each other, each in its own peculiar way, because they share the same dignity.

Leibniz does not locate the uniqueness of each leaf in a "primitive thisness" separate from all its observable properties. Instead, all those properties, taken together, are what he means by the individuality of the leaf. Moreover, these properties do not exist in isolation from those of other leaves and of the rest of the world. As was just noted, each leaf has a place in the seamless fabric of nature. The leaf's individuality is its "complete notion," meaning the total of all the possible things that can be said of it, including everything it ever is, was, or will be. Each leaf, in its unique way, thus reflects the course of the whole universe.

This exalted vision may distract us from an important shift. True, each leaf is unique, but not because of some incommunicable "thisness" hidden beneath its observable properties. Leibniz quietly removes that deeper ground of individuality and replaces it with something fundamentally collective in character. Individuality rests not selfishly "within" the leaf but in its interrelationships with other beings in the cosmic community. The leaf is unique in the way that all beings are, so that each owes its individuality to all the others, never really claiming anything for itself alone. Equality and individuality for all may require sacramental "thisness" for none. If so, the door stands open to even more radical kinds of equality, for nothing grounds the individuality of each in itself, apart from others. Though Leibniz certainly did not intend it, removing primitive thisness will later open the possibility to a thoroughgoing equality of atoms.

Leibniz's own political views were conservative, but one can compare his view of individuality with modern notions of personhood extended to all, not restricted to a privileged few. Holding such views in a democratic age, we tend to grant each snowflake its uniqueness, as if that individuality were the birthright of each. We have ceased to wonder whether the equality of the snowflakes (or people) does not contradict their essential inequality and uniqueness. Despite our passion for equality, the notion of divine providence remains compelling. How can God watch over the fall of a single sparrow unless He can always pick it out from all others?

This sense of essential distinguishability was also crucial for the projects of modern physics, of which Leibniz was one of the founders. He took the argument that led from divine personhood to angelic uniqueness and extended it to the distinguishability of every portion of matter. In the process, he also developed a new conception of space and time, which he considers not as independent entities but rather as expressions of all the possible relations between all conceivable beings. Here he took issue with Newton's conception of space and time as absolute entities, independent of all matter. For Newton, this absolute framework was the essential background of physics, against which each material body had a distinguishable trajectory. Newton showed how extended material bodies could be treated as points endowed with a mathematical quantity of mass or inertia, their individuality utterly distinct. Ludwig Boltzmann, the great nineteenth-century physicist, held that the distinguishability of each mass point was "the first fundamental assumption" of mechanics. Yet if each mass point has no primitive "thisness" and has the same size and mass, only its trajectory in space and time could distinguish it from all the others.

A simple example shows how important this point is. Consider a horse pulling on a stone. According to Newton's Third Law, to every action there is an equal and opposite reaction, so to the action of the horse pulling on the stone there should be the equal and opposite reaction of the stone pulling back just as hard on the horse. Why don't these equal and opposite forces cancel out, making it impossible for the horse ever to move the stone? Many a physics student has puzzled over this seeming paradox and finally learned its resolution: though the forces are equal and opposite, they act on *different* bodies. The force of the horse acts on the *stone,* whereas the reaction of the stone acts on the *horse.* These forces only would cancel if they were to act on the *same* body, which they do not. Moreover, the two bodies are intrinsically distinguishable, and not just by their relative positions: the horse generates enough force to overcome the friction between the stone and the ground as well as the reactive force from the stone, which experiences no other force beside friction, gravity, and the "normal force," by which the ground opposes the downward pull of gravity. In contrast, if horse and stone exerted identical forces on each other, neither would accelerate at all. This example shows that it is absolutely crucial that one distinguish each body from every other. Having labeled each one, Newton's Second Law takes the forces between bodies (presumed to be known from other experiments), their measured positions, masses, and velocities, and then calculates their future trajectories, or even their past trajectories, if desired. If all these pieces of information were perfectly known, it seems that the system of bodies would be perfectly determined for all time. This, then, is how the divine intelligence watches the sparrow. Each bit of matter must have a unique identity to allow the possibility of this perfect determination.

The application of these ideas to chemistry required rethinking the bare notion of "atom." Like the atoms of Democritus and Epicurus, Boyle's atoms lacked any qualities other than size and shape that would explain the variations between chemical substances. Gradually, the concept of "elective affinity" emerged, later articulated as the different valences by which elements could combine, though not yet expressed in terms of electrical forces. As pure forms of various elements were isolated, it was noticed that these elements always combined together to form compounds in the ratios of simple whole numbers. Consider a certain mass of carbon. A certain mass of oxygen will combine with it to make carbon monoxide. Exactly twice that mass of oxygen is needed to make carbon dioxide. Such precise ratios were hard to understand if the elements were continuous, infinitely divisible, for then one would expect that any proportion of elements could be combined, not just a specific ratio. If, however, one atom of carbon combines with two atoms of oxygen to make carbon dioxide and with one atom of oxygen to make carbon monoxide, then the ratio $2:1$ would have a clear explanation.

This concept of "multiple proportions" was a powerful empirical argument, and it moved chemists to adopt the atomic theory because it could account for experimental observations in a way that theoretical arguments about divisibility could not.

However, when John Dalton put forward his atomic theory in *A New System of Chemical Philosophy* (1808), his work was first ignored and then poorly received. Humphrey Davy, the most important chemist of the day, rejected Dalton's atomic theory, though he accepted multiple proportions. To be sure, Davy wrote in 1802 that matter was composed of "particles or minute parts, individually imperceptible to the senses," following in the tradition of Newton and Boyle. In Davy's view, the problem was that those minute particles could not simply be identified with the chemical elements, which Antoine Lavoisier had defined as substances that could not be reduced to simpler forms by chemical means.

By leaping to the conclusion that each kind of atom defined one and only one element, Davy thought that Dalton was being naïve and premature. After all, the great Newton held that one should not traffic in mere hypotheses that tried to go beyond observable laws, such as the law of multiple proportions. Davy and many others shared Newton's distrust of such hypotheses, such as Dalton's atoms seemed. They took Newton's law of gravitation as their exemplar. Even in 1833, John Herschel compared Dalton to Johannes Kepler, who guessed the simple laws of planetary motion but only had an erroneous hypothesis to explain them. Despite his admiration for Dalton, Herschel still awaited the Newton of chemistry, who would find its true mathematical form.

At every stage of the history of science, it is important to enter sympathetically into the reservations and arguments brought against theories that later triumphed. This is not only to show us that great thinkers have their limitations but to let us reconsider the real strengths of the counterarguments that stood against them. Davy thought that Dalton took too literal and simplistic a path from multiple proportions to separate species of atoms for each element. This hypothesis might be sufficient to explain multiple proportions, but it neglects the possibility of other, more sophisticated relations between the minute particles and their compounds. Davy wanted to see all chemical substances in a unified way, rather than broken up into dozens of quite unrelated elements, each with its own kind of atom. To him, Dalton seemed to give up on the quest for a single master element underlying all the rest.

Thus Davy held that Dalton was offering a simple-minded explanation that might impede deeper insight. What is more, Davy thought that Dalton neglected the possibility that the different arrangements of the particles might be crucial, as well as the forces between them, which would negate simple-minded atomism. These considerations about the nature of microscopic forces will return later. Davy also entertained a possibility that bears on the individuality of atoms. "When the particles are similar, the bodies they constitute are denominated simple, and when they are dissimilar, compound." That is, chemical compounds are composed of dissimilar atoms, whereas uniform atoms compose what we call an element. If so, much depends on the different possibilities of similarity or dissimilarity of atoms.

For his part, Dalton offered a compelling argument that all atoms are perfectly alike.

Whether the ultimate particles of a body, such as water, are all alike, that is, of the same figure [shape], weight, etc. is a question of some importance. From what is known, we have no reason to apprehend a diversity in these particulars: if it does exist in water, it must equally exist in the elements constituting water, namely, hydrogen and oxygen. Now it is scarcely possible to conceive how the aggregates of dissimilar particles should be so uniformly the same. If some of the particles of water were heavier than others, if a parcel of the liquid on any occasion were constituted principally of these heavier particles, it must be supposed to affect the specific gravity of the mass [the mass per unit volume], a circumstance not known. Similar observations may be made on other substances. Therefore we may conclude that *the ultimate particles of all homogeneous bodies are perfectly alike in weight, figure, etc.* In other words, every particle of water is like every other particle of water; every particle of hydrogen is like every other particle of hydrogen, etc.

That is, if the individual atoms of hydrogen were not perfectly alike, the element hydrogen would no longer be identifiable as a single element. In Dalton's example, if one happened to collect a certain volume of the "heaviest hydrogen" atoms, they would have a perceptibly different weight than an equal volume of the "lightest hydrogen." What is more, one could get any number in between these two extremes by mixing the number of the "heaviest" and "lightest" or by including some hydrogen atoms of intermediate heaviness. The net result would be that hydrogen would not have a well-defined, precise mass per unit volume, but rather a range of values. This is not unimaginable, but, as Dalton observes, it simply disagrees with the facts: hydrogen does have a precise value for this quantity, not a range of values. The same argument would apply to oxygen (or to any other element) and then also, by extension, to all compounds, such as water.

Dalton seems satisfied with the picture that each atom of hydrogen is exactly like every other. He does not stop to think about how they could manage to be so perfectly alike in every respect. Applying Newton's Laws at the atomic level would seem to require distinguishing every atom, but this becomes practically impossible as one starts to consider more and more atoms. Visible matter contains vast numbers of atoms (for instance, there are roughly 10^{25} molecules in a cubic meter of room air, 1 followed by 25 zeros, or 10 trillion trillion). Are all of them absolutely distinct individuals, each with its own private history and destiny? This is a daunting thought, even for an extreme partisan of individualism, but Newton's Laws require it, strictly speaking. The mind boggles at such large numbers, but presumably an infinite mind should be able to deal with them. Newton himself realized that the equations of motion even of three interacting bodies were not soluble in finite terms. He treated their mutual interactions as an infinite sum of successively smaller terms that could yield only an approximate solution.

Newton's followers devised methods that could give approximate averages for huge assemblages of bodies, as in the cube of air. These methods of "statistical mechanics" ignored the intrinsic differences between the atoms and just dealt with gross thermodynamic properties like temperature and pressure, which are the net result of innumerable atomic impacts. As such, the atoms are treated like coins whose specific dents or peculiarities are ignored in favor of their equal value in circulation. Like coins, their individual value is minuscule compared to the vast sums they make up, which justifies the statistical method. This approach recognizes that terms like "temperature" or "pressure" have meaning only on the macroscopic scale; a few atoms alone cannot have a pressure or temperature because their impacts are too few to create a macroscopic value that a barometer or thermometer could register. In this way, a divergence emerges between the world viewed on the submicroscopic level of the atoms as opposed to the macroscopic level of our experience.

This divergence assumed a particularly challenging form in the treatment of the master concept of statistical mechanics, the concept of entropy. In contrast to every other physical quantity, entropy is not based on the readings of some instrument. Temperature began with the varying readings of thermometers and pressure began with barometers; energy began with measurements of speed (kinetic energy) and height or compression of springs (potential energy), then began to include energy in the form of heat. But entropy was perhaps the first purely constructed physical concept, meant to measure the degree of availability of energy. For instance, it ought to be possible for the table to pass some of its heat into my cold coffee cup. Energy would still be conserved, just transferred from the table to the cup. But it is a brute fact that this never seems to happen, a fact enshrined in the Second Law of Thermodynamics: heat never passes from a colder body to a hotter without some work being done (for instance, the refrigerator must be plugged in for the heat to be removed from inside it). To measure how much heat is available, Rudolf Clausius invented the concept of entropy (Greek for "transformation") and postulated a suitable mathematical form for it, though he had no instrument to give him direction how to do this. In the earliest discussion of mechanical efficiency and available work (1824), Sadi Carnot had imagined idealized machines that were perfectly reversible and ideally efficient; Clausius used those imaginary, perfect machines to guide the formation of the concept of entropy. Once Clausius had made his *definition,* he could state the Second Law in a mathematical form: the entropy of a closed system increases or at least stays the same.

The concept of force began with direct physical experiences of pushes and pulls, measurable in the speeds of bodies and the compressions of springs. Even so, it requires a difficult mental readjustment to reduce the complexity of physical experience to a purely mathematical form. Entropy requires a further adjustment still, for it no longer appeals to any primitive feeling of pushes and pulls. Led by Boltzmann, the partisans of statistical mechanics argued that entropy should be understood as a measure of the probability of states, of the relative likelihood of vast numbers of atoms finding themselves in this or that gross configuration, which we then correlate with a certain temperature or pressure. James Clerk Maxwell applied these statistical methods to relate the observable properties of gases to the properties of their indistinguishable crowds of atoms, though each atom presumably has its own distinct trajectory. In this way, entropy connects atomic chaos and the appearance of order on the macroscopic scale. No wonder *we* don't understand entropy, for it comes not from *our* realm but from what we perceive of the averaged impact of the microscopic realm. It is an amphibian concept that bridges two very different worlds, the microscopic and the macroscopic.

There is, then, a certain justice in the fact that, historically, it was through the concept of entropy that there emerged a new perspective into the nature and identity of matter that would later develop into quantum theory. The first signs of this may be seen well before the emergence of quantum theory in the work of Josiah Willard Gibbs, perhaps the first American to have risen to preeminence in modern science after the pioneering period of Benjamin Franklin and Joseph Henry. Not long after the Civil War, Gibbs constructed what still stands as the mathematical structure of statistical mechanics and devised mathematical formulations of great power that illuminated numerous important problems in physics and chemistry. In the process, he had occasion to consider carefully the entropy of a cubic meter of gas. He noticed that if one considered a mixture of two different gases, one would get a value for their entropy that would not agree with the value if the two gases were identical. This was later called Gibbs's Paradox, for it seemed to imply that the entropy of gases is ill defined if one allows the gases to become indistinguishable.

Gibbs never called this puzzle a paradox, for he felt he was able to resolve it satisfactorily. When the gases become totally indistinguishable, the puzzle is resolved if one takes into account that truly indistinguishable entities can't be treated as if they were still distinguishable. Since one cannot tell permutations of identical atoms apart, one cannot count each of them as if they were different. Yet in Newtonian mechanics, particles have distinct individualities. Gibbs noticed that the straightforward answer he would get for the entropy of a gas of identical atoms was too large by a factor exactly equal to the number of indistinguishable permutations of the N atoms. (As shown in the notes, this factor is the product of the numbers from 1 to N, called "N factorial" and written $N! = 1 \times 2 \times 3 \times 4 \times \cdots \times N$.) If he merely divided his initial answer by this factor, then the "paradox" would disappear and the entropy would be consistent, whether the gases were the same or different. But his procedure showed that something was delicate in the counting of atomic probabilities, even if one followed the procedures that had been extremely successful up to then. The perfect structure of Newtonian mechanics and statistical mechanics seemed to have a gap, and that gap turned on a question of individuality.

Here it is helpful to be more precise. When Maxwell and Boltzmann calculated the state of a gas of one kind of atom, they treated the atoms as if they were distinguishable. What this means is that, unlike Gibbs, they maintained that the atoms were *individuals* although *indistinguishable*. This leads to the paradox that Gibbs resolved by correcting for the indistinguishable permutations of the atoms. In contrast, Maxwell and Boltzmann were doing something potentially problematic: treating the atoms as both individual and not individual at the same time. Clearly, they assumed that, even though we did not distinguish the atoms *in our knowledge,* there was some distinction between them in their very being, which persisted even though we ignored it. However, earlier we noted that Leibniz already removed the "thisness" from matter, leaving only its interrelationships to give it individuality. Newtonian physics expressed this by noting the different trajectories of otherwise identical particles, though Leibniz thought space and time could not confer individuality. What is to keep those trajectories distinct if there is no primitive "thisness" attached to them? Though not made explicit at the time, a significant equivocation had developed about the individuality of atoms, or the lack thereof. Are the atoms really individuals to the extent that their individuality doesn't depend on our inability to distinguish them? But how can they sustain such individuality if they lack "thisness" and also any other marker to distinguish them from each other?

To give a precise form to his way of dealing with this difficulty, Gibbs gave careful definitions of what he called the "generic phase" of an assembly of identical atoms, as opposed to the "specific phase" of distinguishable atoms. By generic phase, he meant a specification of the *state* of the atoms, which includes all their possible permutations. As a later chapter will show, such a concept of "state" offers a way to group atomic configurations that can subsume their individual identities. Careful consideration of the fundamental definition of entropy shows that there is not really a paradox; the crux is how one defines the entropy when the number of particles change (see the notes for details). Because he could resolve the disagreement so readily and simply, Gibbs evidently felt that it could not impinge very deeply on the foundations of physics. Here he showed his characteristic prudence and restraint, even conservatism, for he distrusted ungrounded speculations about fundamental questions. Because of his guarded formulations, his approach to statistical mechanics endured the profound changes worked by quantum theory and remains vital and useful today. Yet it is ironic that he had touched the very spot at which quantum theory would makes a radical departure from Newtonian precedent.

6

The Fields of Light

Newtonian mechanics deals with bodies whose individualities are as sharp and distinguishable as impenetrable atoms. Yet the pursuit of the Newtonian project also led to a strikingly different conception of the world in terms of fluid and penetrable beings: fields. Fields give an altogether different view of individuality, for they can merge and interfere where atoms remain separate and distinct. By indicating this radical new possibility, fields point the way to a crucial element of quantum physics.

Even as he completed his great synthesis of the mechanics of particles, Newton realized something unsatisfactory in his basic approach. He had treated gravitation and all other forces mathematically, setting aside temporarily (or perhaps even permanently) the quest for the true cause of those forces. At the end of his *Principia*, Newton argued that it was enough that his mathematics gave accurate predictions for the motions of bodies, without speculating about the hidden sources of force; beyond that, he would "frame no hypotheses," he curtly noted. So great was Newton's triumph that few could criticize his restraint, which liberated physics from abysmal questions about ultimate causes, to which there may be no answer soon, if ever. By delineating and limiting what mathematical physics could do, he disclosed its powers more than its limitations and was acclaimed as its true founder.

Yet Newton's private ruminations indicate that he was dissatisfied with the scope of what he was able to accomplish, compared with his initial ambitions. He was uneasy with some of his own fundamental ideas. Because he refused to give any account to how the force of gravity worked, Newton stated that it effectively acted at a distance, without any intermediary. Somehow, gravity jumps instantaneously across a gap millions of kilometers wide, giving a force precisely proportional to the inverse of the square of the distance. To many of his contemporaries in Europe, this smacked of old notions of occult qualities, of obscure forces acting spookily across great distances. They stopped laughing when they realized how powerful and accurate Newton's predications were; led by Voltaire, they began following Newton, though Continental mathematicians recast his methods into what has become a more familiar notation (ironically, that of his arch-rival, Leibniz). Newton himself continued to brood over the matter. In a letter to his friend Richard Bentley, he admitted that

it is inconceivable that inanimate brute matter should, without the mediation of something else, which is not material, operate upon and affect other matter without mutual contact, as it must do if gravitation, in the sense of Epicurus, be essential and inherent in it. . . . That gravity should be innate, inherent and essential to matter, so that one body can act upon another at a distance, through a vacuum, without the mediation of anything else, by and through which their action and force may be conveyed from one to another, is to me so great an absurdity, that I believe no man who has in philosophical matters a competent faculty of thinking can ever fall into it.

Realizing this, Newton tried to find a way that action might be transmitted over a distance through some sort of medium pervading space, whose pressure would explain the law of gravity. He left most of this work unpublished, for, as his colleague Colin Maclaurin put it, "he found he was not able, from experiment and observation, to give a satisfactory account of this medium, and the manner of its operation in producing the chief phenomena of nature." Newton also considered Christiaan Huygens's intriguing idea that light was a state of vibration of such a medium, called the "ether," considered to be packed with small, hard particles. When a body began emitting light at some point, those ether particles would be agitated and would transmit their agitation, much as a pool table packed with billiard balls would transmit an impulse imparted to some balls at the edge of the table. Light, then, might be a state of agitation of this close-packed medium, and so too might gravity. But Newton rejected this, for "a dense Fluid can be of no use in explaining the Phaenomena of Nature, the Motions of the Planets and Comets being better explained without it. It serves only to disturb and retard the Motions of those great Bodies, and make the Frame of Nature languish . . . so there is no evidence for its Existence, and therefore it ought to be rejected."

Newton believed his predictions of planetary motion excluded a dense medium filling space, which would perceptibly slow the course of the planets beyond their observed times. However, he continued to consider a "much subtler Medium than Air, which after the Air was drawn out remained in the Vacuum." This "subtle" medium was not close-packed and had void spaces; it would give way before the passing planets and not disturb their orbits. Its varying density would explain the refraction of light and the transmission of heat. In contrast, Newton noted that Huygens's ether "fills all Space adequately without leaving any Pores, and by consequence is much denser than Quick-Silver or Gold." Yet in avoiding one problem, Newton came full circle back to his initial difficulty. If the ether particles do not touch each other (as Huygens had supposed), how then do they interact? Would they not have to act at a distance, the very thing that Newton found absurd?

This problem not only haunts Newton's theories of gravity and light but also comes back to the most ordinary mechanical interactions. Imagine reaching out to touch some object. Does the sensation of hardness or softness mean that my atoms are really touching its atoms? If atoms can exert force only at a distance, the repulsive force they exert will become ever greater as one atomic center comes closer to another. In principle, the force that keeps two impenetrable atoms from occupying the same space becomes infinite as the centers coincide. Thus my atoms do not really "touch" the atoms of any other body but by reaction experience their forces at a distance: the harder the body, the more intense the repulsive force. Even if one tries to fill the space between atoms with a medium that has void spaces, the same paradox will always return in the explanation of the transmission of force between particles of the medium: no two bodies have ever truly touched, despite all appearances.

The very success of action at a distance seems to doom it never to find any explanation beyond itself. Newton seems to recognize this in his letter to Bentley when he seeks "the mediation of something else, *which is not material. . . .*" Here he anticipates the resolution of these paradoxes in the concept of a field, which is not itself material but acts as a mediator between atoms, transmitting force between them. Thus understood, the field was largely the discovery of Michael Faraday and James Clerk Maxwell in the course of their separate investigations of the nature of electricity, magnetism, and light. Their work also rested on experimental discoveries not known to Newton that showed the wave nature of light.

The study of waves was in several ways the source of the idea of fields, and the founder of these studies was Newton himself. In the *Principia,* Newton devoted much attention to "pulses" or waves that travel in a medium such as water. He laid the essential groundwork by showing that the speed of such waves could be expressed mathematically in terms of the density and elastic force of the medium, so that ultimately the motion of waves expresses the microscopic interactions between the atoms of the medium they traverse. He noted the characteristic features of wave motion, especially how waves appear to travel in crooked paths and bend around obstacles. Water waves appear in the still water behind a dock and sound waves are heard even behind an intervening hill. This bending is in sharp contrast to the motion of particles, which do not bend around obstacles.

Newton considered Huygens's suggestion that light also was a wave, but dismissed it because "Light is never known to follow crooked Passages nor to bend into the Shadow." For instance, he noted that the light of the stars never seems to bend around planets that pass between them and us. Though it came much later, Thomas Young's discovery in 1801 that light does in fact bend around obstacles gave decisive evidence for its wavelike nature, exactly fulfilling Newton's criterion. Furthermore, Young observed that light passing through thin slits would interfere with itself, causing characteristic bands of brightness or darkness where the light waves add up constructively or cancel out destructively (which we will return to in a later chapter). These discoveries seemed to doom Newton's own preferred theory that light was a stream of particles. Because of this, Faraday and Maxwell began to consider the nature of electricity and magnetism with new eyes.

Newton recognized that the "most subtle spirit" of electric phenomena cried out for his analysis, and he must have longed to bring his powerful methods to bear on the problem. In the final lines of the *Principia,* he regretted that "there is not a sufficient number of experiments to determine and demonstrate accurately the laws governing the actions of this spirit." He already knew that electrically charged bodies can both attract and repel, whereas gravity only attracts. He would have been intrigued (and perhaps not surprised) to learn that the force experienced by charged bodies obeys an inverse square law exactly parallel to that of gravity. He may have surmised that the deeper reason for this striking parallelism has to do with the three-dimensionality of space, in which conserved quantities like charge or mass are only consistent with inverse-square forces. However, for him these would have been groundless speculations, lacking the necessary experimental evidence.

As he uncovered that evidence, Faraday gradually elaborated a new view of electricity and magnetism in terms of what he called *lines of force* and also *fields,* terms to which Maxwell gave mathematical expression. The contrasting stories of the two men illuminate the concepts that they shaped. The son of a blacksmith, Faraday began working at age twelve as a bookbinder's apprentice. He read the books he bound, especially those about science, and approached Humphrey Davy after hearing his lectures at the Royal Institution in London. It has been said that Faraday was Davy's greatest discovery, though there were moments when Davy was a difficult and jealous patron. From these humble beginnings, through many years of reading and ceaseless experimentation, Faraday became a great luminary of European science. He hated the term "physicist," which had only recently (1830) been coined by William Whewell, and wanted to be simply a philosopher, and an unmathematical philosopher at that. Writing to the eminent mathematical physicist André-Marie Ampère, Faraday confessed:

I am unfortunate in a want of mathematical knowledge and the power of entering with facility into abstract reasoning; I am obliged to feel my way by facts closely placed together so that it often happens I am left behind in the progress of a branch of science, not merely from want of attention, but from the incapability I lie under of following it, notwithstanding all my exertions. . . . I fancy the habit I got into of attending too closely to experiments has somewhat fettered my power of reasoning, and chains me down; and I cannot help, now and then, comparing myself to a timid ignorant navigator who, though he might boldly and safely steer across a bay or an ocean by the aid of a compass which in its action and principles is fallible, is afraid to leave sight of the shore because he understands not the power of the instrument that is to guide him.

Despite his sincere modesty, there may also be some irony in his tone, for it was his extraordinary attention to stubborn experimental facts that led him to his great discoveries, even though he lacked mathematical sophistication or perhaps just because of that. Faraday was the most practical of men, intensely attentive to the vivid detail of phenomena. He was a virtuoso of experiment, insightful, indefatigable, and endlessly inventive. Yet though he grew up with notions of electricity and magnetism as palpable stuff, this hard-headed man convinced himself that the true reality was made up of invisible lines of force, not the charges or magnets that were usually regarded as their sources.

Consider, for instance, a magnet sprinkled with iron filings whose alignment seems to outline invisible lines of force (figure 6.1). Faraday gradually took these lines more and more seriously, arguing that they are present even before the iron filings indicate their direction. Finally, he turned the ordinary picture on its head; instead of the magnet generating its lines of force,

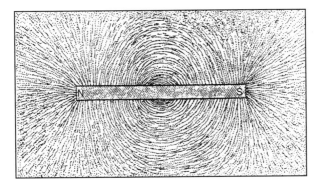

Figure 6.1
Faraday's illustration of the lines of force around a magnet (*Experimental Researches in Electricity*, 1855).

he suggested that "magnets may be looked upon as the habitations of bundles of lines of force." Likewise, he proposed that electric charge is better understood not as a material substance but merely as the apparent locus of converging electric lines of force. Here Faraday considered the radical vision of the eighteenth-century Dalmatian natural philosopher Rudižger Bosković (Roger Boscovich), who emerged in the first generation after Newton. Boscovich replaced Newton's "hard, massy" atoms with massless "points of force," mathematical centers devoid of matter yet each the locus of Newtonian forces. By so doing, Boscovich thought to advance beyond Newton by reducing all atomic forces to a single master force between these mathematical points. Whatever his relation to Boscovich's ideas, Faraday wished to retain an open mind about Dalton's atomism. For his part, Faraday thought it preferable to solve the duality of matter versus force by giving priority to the deeper and more intelligible of the two, as he saw it. In his vivid reconception of the world, lines of force are the true reality behind "matter."

Faraday's daring vision persuaded Maxwell, who took it as the basis of his mathematical theory of electricity and magnetism, though he did not completely follow Faraday in rejecting matter in favor of the field. Born into a totally different milieu than Faraday's, Maxwell was the son of a wealthy Scottish family and was a brilliant mathematician who excelled at Cambridge. Yet his greatest work was to turn Faraday's vision into beautiful equations, and he always acknowledged his debt. Though he admitted that Faraday could not understand a single page of mathematical symbols, Maxwell insisted that Faraday's way of looking at lines of force shows Faraday "to have been in reality a mathematician of a very high order—one from whom the mathematicians of the future may derive valuable and fertile methods." In an admiring letter, Maxwell wrote Faraday that "you are the first person in whom the idea of bodies acting at a distance by throwing the surrounding medium into a state of constraint has arisen, as a principle to be actually believed in." For Maxwell, Faraday's lines of force become a "state of constraint of the surrounding medium," which is thrown into a state of polarization that Maxwell calls a field. Newton's tentative thoughts about fields as immaterial mediators find their realization in Maxwell's equations, which he modestly described as a mathematical "translation" of Faraday's ideas.

Yet without mathematics, in his own way, Faraday realized that light was "a high species of vibration in the lines of force which are known to connect particles." However, Faraday's turn of phrase reveals a significant difference. Maxwell considered it logically necessary that fields travel in a material medium that he called the ether, just as water waves need water. Faraday was bolder; he thought "to dismiss the ether, but not the vibration." Faraday imagined lines of forces vibrating in a vacuum,

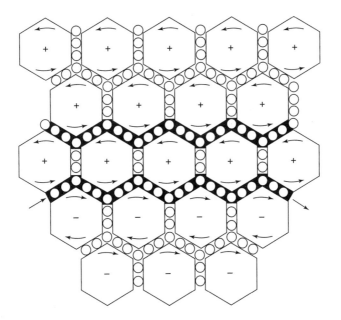

Figure 6.2
Maxwell's imaginary gear mechanism ("On Physical Lines of Force," 1861).

like an invisible spider web whose tremors are light itself. Here Faraday anticipated Einstein's argument that there is no material medium in empty space; despite our intuition, light waves do not require any palpable substance to support their passage.

Maxwell initially reached his equations by hypothesizing a mechanical structure in the ether, an elaborate network of gears that he treated with surprising realism in order to depict the coupled behavior of the electric and magnetic fields (figure 6.2). After he reached his mathematical goal, he set aside this mechanical scaffolding, realizing that these visual explanations were only hypotheses that would detract from the generality of the equations themselves. Indeed, his successors forgot his gear wheels. They treated his mathematics without reference to the ether, which seemed more and more problematic, absent any positive evidence of its existence or properties. Maxwell himself noted disturbing paradoxes that haunted the ether; it had to be exceedingly tenuous so as not to impede the planets perceptibly, but at the same time rigid enough to explain the high velocity of light waves.

Faraday was just the man to appreciate a physical approach to these invisible forces. To be sure, he had trouble with the mathematics, writing to Maxwell that "I was at first almost frightened when I saw such mathematical force made to bear upon the subject, and then wondered to see that the subject stood it so well." There is gentle irony here, as well as respect for the power of the mathematical symbols that Maxwell was shaping. Even so, Faraday maintained a certain pride in the integrity of his own views, even as he self-deprecatingly called himself merely a "laborer." He knew the value of his labors, or at least felt confident that posterity would sift the gold from the dross. Yet there is also the note of a wistful Moses, who sees the promised land from afar and recognizes that he will not himself enter into the fullness of it. Though Faraday had already anticipated that light is a vibration of electric and magnetic lines of force, he could not calculate the speed of these vibrations. For this, more was required than a qualitative picture, however inspired its physical insight might be. The triumph of Maxwell's mathematics was to show that the speed of traveling electromagnetic waves agreed exactly with the observed speed of light.

In their different ways, the approaches of Faraday and Maxwell illuminated the novel character of the field. Faraday was right to consider the field both physically real and yet immaterial. Maxwell was right to devise a new mathematics to trace this invisible voyager, though he was wrong to think it required a physical medium to support it. In both cases, the field is perplexing because its nature is very different from that of a particle. At least as Newton conceived it, the distinction between the individualities of two particles is so marked that it is impossible for them ever to coincide or for either of them to alter the being of the other; the particles might collide and rebound endlessly without ever becoming confused as to which of them is which. The situation is very different with light, which can interfere with itself and thus cause spots of greater or lesser intensity, even of total cancellation, as Young discovered. In the case of material bodies, Maxwell notes that "we cannot suppose that two bodies when put together can annihilate each other; therefore light cannot be a substance." Yet it appears that "one portion of light can be the exact opposite of another portion, just as $+a$ is the exact opposite of $-a$, whatever a might be." Since no substance can exist as a negative, Maxwell concludes that "light is not a substance but a process going on in a substance," meaning the ether.

Waves have a curious kind of being, even water waves. A wave in deep water, moving in a certain direction along the surface, is composed of water that itself never actually moves in that direction. Instead, the water molecules only go up and down, as a buoy will reveal, while the wave appears to move horizontally. Even though it appears to be as solid and substantial as any material object, a wave is a moving form, not simply a single body. At each successive instant, a wave is an ever-new assemblage of water, whose changing individuality is subsumed under the outward form of the whole wave. In this sense, a wave is a *process,* not a substance, as Maxwell noted. Waves interfere with each other because they are interchangeable and thus not distinguishable; two processes can coincide in space and time, but two substances cannot. Thus the wave reveals a whole new possibility of identity, for one identifies waves by their amplitude (that is, height) and by their wavelength or frequency, rather than by the ever-changing bits of water that make up the wave at any time. Even if one distinguishes one wave from another as they approach from a distance, when they pass through each other, one can no longer sustain that distinction, which makes sense only in the context of ordinary bodies.

Faraday extended this concept to the invisible fields that constitute visible light, understood as waves in the lines of force. He was deeply struck that these vibrations were not fixed in place but were curved and traveled through space and time, which he took as touchstones for their physical reality. He saw "a motion, not a form." Stationary fields surround static charges or magnets, but when these sources accelerate, their fields leap free, manifest as waves of light traveling through boundless space. Whether stationary or moving, atoms give rise to fields that are completely indistinguishable. This bears on the individualities of the atoms themselves, for Faraday's radical view had a further consequence, though one not drawn at the time. If charge and matter do not really exist except as a way of speaking about the density of field lines, which are indistinguishable in themselves, then material particles are not really distinguishable either. Faraday's vision of a world of fields put in doubt the kind of individuality appropriate to impenetrable particles.

The concept of a field is elusive. On the one hand, it is a mathematical abstraction that specifies a magnitude and a direction at every point of space; for instance, an electric field specifies at every point the magnitude and direction of the electric force that would be experienced by an imaginary test charge placed there. Thus the field is a web of numbers, each a generalized, unindividualized magnitude. On the other hand, Faraday endowed the field with ultimate physical reality, as if it were a substantial entity. If both of these conceptions are maintained, the field must then blend the abstract quality of mathematics with the concrete reality of discrete matter. As will emerge, this will show itself particularly in the way the lack of individuality of fields transforms the separateness of particles.

Maxwell considered that the infinite hardness and impenetrability of atoms (which Newton considered axiomatic) was a "vulgar opinion" derived only from commonsense experience and hence not necessarily valid on the atomic scale. Moreover, he notes that molecules and atoms must vibrate in order to produce the spectra of the various elements. Such vibration is incompatible with hard, rigid atoms or with massless centers of force. Accordingly, Maxwell did not follow Boscovich's vision but envisaged new possibilities. He returned to Dalton's argument that the atoms of each species must be identical, at least in observable properties. Extending it beyond the sphere of earthly laboratories, Maxwell argued that "there are innumerable other molecules, whose constants are not approximately, but absolutely identical with those of the first molecule, and this whether they are found on earth, in the sun, or in the fixed stars." Writing in 1872, in the aftermath of Charles Darwin's *Origin of Species* (1859), Maxwell consciously made the contrast with Darwinian evolution, in which there are "intermediate links" between species:

In each [biological] species variations occur, and there is a perpetual generation and destruction of the individuals of which the species consists. Hence it is possible to frame a theory to account for the present state of things by means of generation, variation, and discriminative destruction. In the case of the molecules, however, each individual is permanent; there is no generation or destruction, and no variation, or rather no difference, between the individuals of each species.

There is no "theory of evolution" for atoms, nor can there be, Maxwell concludes, and he expresses renewed astonishment at the exact equality of atoms, whether found in earthly coal or in a meteorite that falls to Earth.

Here Maxwell adds "a fact of a different order": atomic periods of vibration are also exactly equal, since the same spectral patterns are observed on Earth as in the light of distant stars. The exact sameness of atomic vibration is even more surprising, for it depends on the dynamic qualities of atoms in motion, not just their mass. Maxwell notes that no physical principle he knows could prevent vibrating atomic charges from emitting *any* color of light over a continuous range of values, given the proper conditions. Surely the variation in conditions is sufficiently great between Earth and a distant star; though he does not explicitly say so, it seems inexplicable that there are discrete spectra at all. Furthermore, Maxwell's contemporaries were beginning to discover the true vastness of the cosmos and he was accordingly aware that, as one looked out at the stars, one is looking backwards in time. The spectra of those distant stars show that atoms had the same properties eons ago, when the light that we now see left the stars.

Although Maxwell's usual tone is commonsensical, at this point he is notably awed. "But in the heavens we discover by their light, and by their light alone, stars so distant from each other that no material thing can ever have passed from one to another; and yet this light, which is to us the sole evidence of the existence of these distant worlds, tells us also that each of them is built up of molecules of the same kinds as those which we find on earth." Each molecule is thus stamped with a "royal cubit," a mark of such uniformity that gives it "the essential character of a manufactured article, and precludes the idea of its being eternal and self-existent." The image of manufacturing is striking, yet apt, given the importance of the industrial revolution in his time. Maxwell discerns the hand of the great Manufacturer in His wonderfully uniform products.

Maxwell realizes that here we are "very near to the point at which Science must stop" because it has reached beings that have "not been made by any of the processes we call natural." He realizes that he stands at the verge of creation. "Science is incompetent to reason upon the creation of matter itself out of nothing. We have reached the utmost limit of our thinking faculties when we have admitted that because matter cannot be eternal and self-existent it must have been created." In identical atoms, he discerns the "ineffaceable characters" that "are essential constituents of the image of Him who in the beginning created, not only the heaven and the earth, but the materials of which heaven and earth consist." Here his discourse ceased, as befits his recognition of the Maker's incomparable power. With wonder, Maxwell avowed that it is "the peculiar function of physical science to lead us to the confines of the incomprehensible." At the same time, Maxwell opened the door to new consideration of the strange "identicality" of atoms.

7

Entanglement

At the end of the nineteenth century, particles and fields were the two guiding concepts of physical theory. A thoughtful observer might have been disturbed at their deep difference, as if two antithetical concepts could both be at the heart of things. For his part, Faraday resolved this paradox by making the field supreme. In his view, what we call a particle is just the way fields look in certain circumstances. The immense success of Maxwell's equations could be taken as confirmation of this view, though he did not follow Faraday in reducing matter to being ultimately fields. These elegant equations encompassed all known electric and magnetic phenomena and predicted the speed of electromagnetic waves even before Heinrich Hertz measured it and found it equal to the speed of light. Maxwell's equations describe the interconnection of the electric and magnetic fields, which in turn guide the motion of charged particles. From Faraday's viewpoint, the time was ripe for experimental confirmation that charge was nothing but an intense nexus of fields.

It was, then, a considerable surprise that careful experiments gave powerful evidence that electric charge came in elementary units, which have every property of particles. These experiments began with the discovery of "cathode rays," luminous electric discharges in evacuated glass vessels, the ancestor of television and video monitors. Benjamin Franklin demonstrated this phenomenon, but the possibility of careful experiments waited on the development of pumps capable of creating a high vacuum, without which the cathode rays collide with the residual gas in the tube, making only a diffuse glow, not a workable beam. Such pumps were available after about 1850; in the 1880s, William Crookes showed that the cathode rays cast shadows and can turn a tiny paddle wheel. Moved by the eerie glow in his tubes, Crookes thought he had discovered the ectoplasm described by spiritualists. Though the English physicists (following Crookes) argued that the rays were composed of tiny particles, the Germans held that they were a kind of electromagnetic wave.

This debate finally was resolved by a series of telling experiments. In 1897, J. J. Thomson succeeded in making cathode rays travel in circles by passing them through crossed electric and magnetic fields. Electromagnetic waves would not have responded to such fields, according to Maxwell, because they have no charge in themselves. In contrast, cathode rays evidently seemed to have inertia and hence mass as well as charge, shown by the deflection of their paths under electric or magnetic force. Maxwell's equations allowed Thomson to measure their ratio of charge to mass, which he found to have a single, universal value, easily understandable if cathode rays really were streams of particles, each of which had the same charge and mass. Thomson fortified this argument by making similar experiments in 1899 on the particles produced when light hits a metal plate (the "photoelectric effect" later treated by Einstein, the basis of the "electric eyes" that open doors and trip alarms). He was the first to show that these particles have the same charge as the cathode rays, both now called "electrons" (a term George Johnstone Stoney coined in 1891).

By 1910, Robert Millikan devised an ingenious experiment that measured the charge of the electron more precisely and found it to be unique and universal, as Thomson had anticipated. Millikan took a tiny oil drop, charged by static electricity as it emerged from an atomizer. He used electric fields to move the drop up and down, timing its free fall under gravity and comparing it with the time it took to rise when sufficient electric field was applied. He did this with thousands of drops of varying size and charge; he used radioactive sources to cause changes in the charge of the drop, which were always a simple multiple of the electron's charge.

It became very difficult to see how discrete electrons themselves could be explained in terms of continuous fields. Indeed, Maxwell had cautiously retained massive particles within the framework of his field equations. Yet any return to particles would take place under the shadow of fields; in the realm of light, the characteristic wave phenomena of interference were experimental realities that confirmed Maxwell's field equations. As so often in the history of science, paradox was the midwife of a new vision. By a further irony of history, the most disturbing innovation of modern science was introduced by a champion of the older physics. The story of Max Planck illustrates how new ideas can emerge even against the wishes of their discoverer. Once he realized the magnitude of his innovation, Planck did everything he could to get around it or reverse it. His loyalty to the older physics did not, in the end, stop him from recognizing the disturbing truth of what he had first suggested, though it was other, younger people who really carried his idea to full fruition. Albert Einstein stated the quantum idea in 1905 with a fuller sense of its shock, readier than Planck to embrace its radical quality. I hope what follows will help clarify the true dimensions of what Planck did, without ignoring his resistance and denial. In the early phase of a new idea, even the discoverer may not fully realize what has been discovered.

From his youth, Planck's imagination was fired by a quest for the absolute, for truths that would transcend what he called the anthropomorphic limitations of human thought. He wanted truths that would not depend on local prejudices, or human foibles, but would be as true for Germans as for Americans, for Martians as for Earthlings. Like Einstein, he sought a refuge from what is merely personal, from the nightmare of history and the tumult of human confusion. He found such truths in what now is called classical physics: in Newtonian mechanics, in Maxwell's equations, and especially in the sweeping generalities of the laws of thermodynamics, where his own quest began. Yet these laws seem to have different characters. How do the laws of thermodynamics relate to Newton's mechanics and to Maxwell's equations? Is thermodynamics perhaps just a consequence of classical mechanics? Or is it somehow different? Planck started by trying to derive the laws of thermodynamics from those more basic principles of mechanics and electrodynamics.

His early attempts came up against devastating criticism by Ludwig Boltzmann. With characteristic vehemence, Boltzmann argued that there was no way Planck could succeed. The Second Law of thermodynamics is above all a statement of *irreversibility;* in general, processes involving heat are irreversible, as when a coffee cup resting on a table loses heat and is unable to draw it back. The only imaginable exceptions would have to use one of Carnot's imaginary heat engines that are specifically designed to be perfectly reversible; however, in the ordinary world these idealized stipulations fail and heat moves irreversibly. On the other hand, Newton's Laws and Maxwell's equations are always *reversible* in time; if they allow some process to occur, they must allow it also to happen in reverse. Thus, Boltzmann noted, there is a profound difference between thermodynamics and mechanics; one is irreversible, the other reversible, and Planck's simple idea of reconciling them was bound to fail.

Boltzmann thought he had solved this problem; he argued that thermodynamics should be understood in terms of probability, and that apparent irreversibility arose not from the laws of mechanics but from peculiarities of the *initial conditions,* the ordered starting points of familiar processes. For instance, initially the heat is confined in the coffee cup. As time passes, there is nothing in the reversible laws of mechanics that can stop the heat from creeping back into the coffee cup, except that this would be very unlikely to happen. It is overwhelmingly more likely that the heat continue to diffuse away from the cup into the table than that (through a long and improbable chain of events) the atoms in the table transfer their energy back into the cup. Boltzmann exalted the concept of probability as the key to understanding the apparent irreversibility of the world: from this perspective, the Second Law of thermodynamics is not absolutely true, but only extremely probable. At any moment, the coffee cup might begin to draw heat from the table, but, in all likelihood, I will have to wait far longer than the age of the universe for that to happen.

Planck resisted Boltzmann's idea for a long time, holding to the absoluteness of the Second Law. After all, no one had ever seen the Second Law violated, at least on the scale of coffee cups. Likewise, no one had noticed a violation of the law of gravity; nothing had ever fallen up. Isn't this sufficient justification to consider both gravity and the Second Law to be absolute? Yet Planck was moved by Boltzmann's general approach, because it relied not on human-oriented concepts like "available work" but on a far less anthropomorphic concept of probability. Though he suspected probability of being incompatible with absolute truth, Planck saw in Boltzmann's statistical mechanics a powerful way to approach physics in great generality. He also recognized that Boltzmann's arguments against him were irrefutable, and so he capitulated.

Yet both men wanted to hold on to the absoluteness of the Second Law, despite the possibility (however remote) of its violation. To do so, Boltzmann went beyond his initial assertion that probability is the basis of entropy. Though now he is generally remembered for this assertion, he also formulated a less well-known "principle of elementary disorder," which states that the atoms are always in a disorderly state, that they can never conspire to organize themselves. To be sure, it is hard to imagine how atoms could ever organize themselves, lacking any awareness or volition. Yet Boltzmann here tried to exclude the possibility that mere chance might bring about an apparently ordered state; he did so to avoid the violations of the Second Law that those fortuitously orderly states could cause. Logically, at least, this principle would be sufficient to rule out large-scale violations of the Second Law because it forbids the appearance of the necessary small-scale correlations. However, this principle would block the occurrence of certain patterns allowed by Newtonian mechanics, for mechanically there is nothing to exclude *any* possible state of the atoms, however orderly. Boltzmann did not notice that this principle, taken seriously, would conflict with the full generality of Newton's Laws. It came back to haunt Planck.

Later in his life, Planck wryly remarked that novel scientific theories do not so much convince their opponents as outlive them; the old guard eventually dies out and the young people accept the innovations. Despite the acuteness of his observation, Planck was himself a counterexample, for he took up the cause of his former nemesis with surprising enthusiasm. He neither entrenched himself in self-defense nor gave up in despair, but bravely took up Boltzmann's new methods and returned to his project. To study the operations of the Second Law in its most general context, Planck treated the problem of an oven heated to high temperature. This idealized oven was supposed to be perfectly black inside, meaning that it would absorb light of any color that entered it. Earlier physicists had already shown that when such an oven was heated it would emit a certain spectrum of colors depending on its temperature, but quite independent of its size, shape, or even its material. The spectrum of an ordinary incandescent light bulb gives a fair approximation to the light coming out of a hole in such a black oven; the Sun's spectrum is quite close to that of a black body heated to about 6000 degrees centigrade. But what then determines the exact properties of this "black-body radiation" (as it is called), if indeed it is independent of all material properties of the body itself? Planck thought that solving this problem would show the way to whatever absolute quantities determine the spectrum of light inside the box. In effect, he treated the light inside the oven as if it were a kind of gas, and then applied Boltzmann's methods of statistical mechanics to it. He knew also that his colleagues were measuring the spectrum of black-body radiation with new accuracy (including the previously unobserved infrared and ultraviolet parts of the spectrum) and hoped to test his theory against their observations.

When Boltzmann had applied his ideas to an ordinary gas, he found it convenient to pretend at first that each atom could gain or lose energy only in discrete amounts, as if they could only exchange coins of a given value. At the end of his calculations, Boltzmann would let this minimum energy transaction go to zero, indicating that the atoms really could exchange any amount of energy in a continuous gradation, not just discrete coins. However, when Planck applied an analogous argument to his "gas" of light and let the exchanged energy become continuous, he noticed that the theory disagreed sharply with the experiments. He soon found that his theory would agree with the experiments only if he stipulated that the exchange was always of a discrete amount of energy, equal to the frequency of the light times a certain constant, h (now called Planck's constant). What followed Planck later called "an act of desperation." The experiments pointed to a small but quite definite value of h, meaning that energy could *not* be continuously exchanged between the atoms. Such a discrete bundle of energy Planck called a *quantum*.

Planck was perplexed, but honest enough to announce in 1900 what he had found. Privately, he told his oldest son that he had found something as significant as the discoveries of Newton, but his emotion was anxiety, not exultation. He was deeply disturbed that what he found made no sense to him. At first, his discovery made no great stir, and he had some years to try to make sense of it. During that time, Einstein did notice Planck's work; indeed, while Einstein was still working at the Swiss Patent Office, Planck was the first influential person to notice his work on relativity, bring it to publication, and draw attention to it. In one of those three seminal papers he published in 1905, Einstein took Planck's idea of the discrete chunk, or "quantum," of energy to a further level of seriousness. Einstein considered the emission of electrons by light shining on metal (the photoelectric effect that Thomson had used to determine the electronic charge). Einstein noted that the light acted exactly as if it were corpuscles of a certain discrete energy that could collide with the electrons and sometimes give them sufficient energy to escape the metal.

Planck must have been struck by this bold extension of his original hypothesis, but he was wary of it. In his own writings, Planck tried to avoid the quantum hypothesis in some way or to limit it as much as possible. Perhaps the light in the oven is *emitted* in discrete energy chunks but can be *absorbed* continuously; perhaps light can interact with matter classically but does something peculiar only after emission. Planck realized that his hypothesis pointed toward the rejection of Maxwell's theory of electromagnetism, which was purely continuous and had no place for energy quanta. He tried everything he could think of to get rid of the quantum, to save the beautiful classical theories. Einstein, in contrast, blithely ignored the havoc his idea wreaked on Maxwell's idea of light as electromagnetic waves.

Planck's impulses tended toward reconciliation, striving to harmonize the new with the old, but he still kept faith with his disturbing idea. He constructed a new derivation of his result, trying to find positive arguments that would go beyond the need to fit experimental data, which had driven his original hypothesis. Beginning with Boltzmann's idea that statistical mechanics is the right way to understand gases, Planck emphasized the central importance of the atom in this approach. This was controversial because there still remained prominent doubters of atomic theory. Such eminent figures as Wilhelm Ostwald and Ernst Mach believed that atoms were an explanatory fiction, a theoretical construct that did not require the existence of real atoms. Mach died in 1916, still unconvinced, though he was shaken when he saw a visible indication of atomic discreteness, the flashes of light emitted by radioactive substances.

Most of the doubters were eventually persuaded by studies of Brownian motion, such as the third landmark paper Einstein published in 1905. In 1807, Robert Brown had noticed the ceaseless jiggling of grains of pollen or other small objects in his microscope. At first, Brown thought this motion might be evidence of a new form of life, but soon he was able to show that the moving particles were quite lifeless, despite their constant jittering. Others suggested that the movement was the result of innumerable collisions with the atoms surrounding an object, which generally result in a small, but perceptible, motion since unequal numbers of atoms hit on different sides. From this, Einstein was able to deduce Avogadro's number, which characterizes the number of atoms in a standard mass.

Though Amedeo Avogadro was a chemistry teacher trying to articulate the atomic theory, his number could not be determined through purely chemical means, which use equivalent numbers of atoms reacting together but have no independent means of measuring their number. That determination waited until physics was applied to the motion of atoms in gases. Maxwell was the pioneer of this kinetic theory of gases and extended the pioneering measurements of Josef Loschmidt (1865) to yield improved values for the size of atoms and Avogadro's number. Independently, Einstein and Marian von Smoluchowski came up with a dozen different ways of determining this crucial atomic constant from such diverse phenomena as the blueness of the sky and the opalescence of gases near their critical points. Finally, Jean Perrin gave these calculations the confirmation of exacting experiments. If atoms were not real, how could all these different methods agree?

However, well before such arguments emerged, Planck was a convinced atomist on the basis of his understanding of the concept of probability. He argued that there could be no statistics without discrete objects to count. That is, a statement of probability gives the relative likelihood of a finite number of discrete outcomes (or discrete classes of outcomes), each one of which is equally likely to occur by itself. Thus the probabilities of rolling several dice rely on the discrete states given by the faces of a single die, each one of which is equally likely to be thrown. In this way, Planck argued that the understanding of physics leads to the concept of entropy, which measures the probability of states, which must then be discrete or atomic. This may have been the first argument for atomic theory since the Greeks that grew out of larger philosophical ideas, rather than specific chemical or experimental evidence.

So far, Planck was just commenting on Boltzmann's approach. At that point he extended that approach to cover not only gases or assemblies of atoms but light itself. Planck was sure that light, even though immaterial, obeys the laws of thermodynamics; those laws seemed so general that he could not imagine an exception to them. If so, then light too must have an entropy, which is a measure of the relative probability of different states of the light. Like matter, light also tends toward states of greater probability and so greater entropy. But that must mean that there is some discrete basis for counting the different probabilities of the states of light; there must be, so to speak, "atoms" of light, which could only mean discrete states of light energy, or "quanta," since light itself is not material.

This simple but powerful argument Planck had found by 1908; as he applied it to rederive his results for the black-body spectrum, he realized that it had a peculiarity buried within it. In order to probe the states of the light in the oven—the "light-atoms"—Planck put an imaginary radio into the oven, tuned to one single frequency. The radio receives a signal whose strength measures the amount of light energy in the oven at that frequency. But in counting the different ways the light-quanta could share the energy, Planck realized that he had to treat the quanta as absolutely indistinguishable. In effect, he came upon that same factor that Gibbs had noticed earlier, the factor that expresses how many different, but indistinguishable, ways a certain state can be formed (a factor of $N!$ for N indistinguishable particles, as shown in the notes for chapter 6). If Planck tried to treat the light quanta as distinguishable, he got results that disagreed with the experiments. Looking back into his earlier work, one sees this same peculiar counting appear as a necessary ingredient for his solution of the black-body spectrum. At that point, Planck did not comment on it, perhaps more struck by the necessity of discreteness, the "desperate" step of letting h be a finite number, not zero. Even in 1909, he presented the strange counting without seeming to realize its strangeness, as if it were a peculiarity of his imaginary radio rather than of the radiant light in the oven. Yet Planck's integrity kept him from denying what he had asserted, even while he struggled to overcome its paradox.

It was not until 1926 to 1928 that the structure of quantum theory was articulated in the form that became central to twentieth-century physics and that stood unchallenged into the following century. That theory, associated with the names of Werner Heisenberg, Erwin Schrödinger, Max Born, and Paul Dirac, triumphed in every test and gave a new level of power in predicting atomic and nuclear phenomena of every sort. Quantum theory gives a mathematical description of the state of a system in terms of "wave functions," which can be represented by vectors in a many-dimensional "Hilbert space." This "space" is not at all the physical space of our experience, but it allows a lucid and beautiful treatment of the mathematics of the theory. Indeed, in Hilbert space quantum theory has a kind of simplicity and naturalness that it loses when one returns to the space and time of our experience, as if it were a duck out of water.

For this reason, quantum theory remained enigmatic and paradoxical. Readers puzzled over accounts of weird mixtures of particles and waves, of "spooky action at a distance" (Einstein's phrase), of uncertainty and indeterminism. The experts were no less challenged to find some intuitive understanding of the strange concepts whose mathematical formulation they used with such power. The theory still resists every attempt to visualize it, for ordinary vision turns out to be inadequate to grasp fundamental realities on the atomic scale. As Niels Bohr argued, our words and concepts are geared to our experiences as macroscopic beings, composed of vast numbers of atoms. However high it flies, our vaunted imagination still relies on its starting point, common human experience. To go where intuition fails, physics relies on logic and on the abstract language of mathematics, which tries to extend the speech of the tribe to a realm beyond the all-too-human.

Even after one has struggled to master the mathematical language of quantum theory, as eminent a physicist as Freeman Dyson suggested that the best that one can do is to say "I understand now that there isn't anything to be understood." Even Richard Feynman, a master of finding simple ways to understand complex ideas, threw up his hands. After giving a classic summary of the basic rules of quantum theory, he confessed that "one might still like to ask: 'How does it work? What is the machinery behind the law?' No one has found any machinery behind the law. No one can 'explain' any more than we have just 'explained.' No one will give you any deeper representation of the situation. We have no ideas about a more basic mechanism from which these results can be deduced."

Yet without denying the truth of these remarks, I think that there is a way of drawing the strangeness of quantum theory to a single, telling point, one that does not dissolve that strangeness but may make it more intelligible. I think that the heart of quantum theory is its radical innovations on the question of individuality.

This insight has been obscured because, in most accounts of quantum theory, the loss of individuality is treated as a consequence rather than as a fundamental postulate. Yet the theme of lost individuality can be traced back to Planck, though he did not emphasize the issue and did not fully grasp its import. In 1911, Ladislas Natanson in Cracow and Paul Ehrenfest in Leyden pointed out that Planck's strange way of counting implies that quanta are not independent; their lack of individuality requires strange correlations utterly different from the behavior of distinguishable classical particles. Here one remembers the field, whose modes interpenetrate and interfere, not having distinct individualities. Einstein's work of 1909 already implied that a full theory of quanta must somehow combine particle with wave. Applying these ideas to matter rather than light, Satyendra Nath Bose and Einstein used arguments of indistinguishability around 1924 to establish the quantum laws of ideal gases. When Schrödinger and Heisenberg wrote down their different (but mathematically equivalent) formulations of the principles of quantum theory, the utter indistinguishability of electrons or of light quanta was implicit in their formalism, though it emerged only as the theory was explored.

Some of these physicists were also aware of larger philosophical issues, and this new kind of individuality struck them. In 1931, Hermann Weyl wrote with some astonishment of two "quantum twins," Mike and Ike: "It is impossible for either of these individuals to retain his identity so that one of them will always be able to say 'I'm Mike' and the other 'I'm Ike.' Even in principle one cannot demand an alibi of an electron!" That is, one can never account for its whereabouts. In 1952, Schrö-dinger emphasized that the issue was "*much more* than that the particles or corpuscles are all *alike*. It means you must not even *imagine* any one of them to be *marked*—'by a red spot' so that you could recognize it later as *the same*."

Returning to the argument about identical leaves, the inge-nious gentleman might have won the bet by presenting the prin-cess with identical electrons instead. It is hard to avoid the conclusion that Leibniz was wrong; some things really are iden-tical, despite all our ordinary experience. Such identical twins must be strange indeed.

8

Identicality

Let me now *begin* with the assumption that electrons utterly lack any individuality, rather than treating this as a consequence of something else. From this fundamental postulate, let us see how the theory then unfolds. All electrons have exactly the same charge and mass as well as spin (which indicates an intrinsic quantity of angular motion that cannot adequately be understood in terms of classical spinning bodies). As Millikan emphasized, this exact equality is not merely statistical. To high precision, there is *no* variation whatever in charge, mass, or spin from electron to electron. In contrast, in classical physics, one would expect a certain spread in these values, if one imagined electrons to be small physical objects. After all, in biology, the different individuals in a certain species differ slightly, though not so much as to pass certain bounds that define the species.

Electrons are all radically *equal*. An electron's individuality *is* its species, and nothing more; it is always an instance of electronhood, never a "noninstantiable" individual. In this, electrons are the opposite of Aquinas's angels, each of whom is the unique representative of its own species. Because of this, the term "particle" is misleading, since it suggests a body with certain characteristic features or markings, or at least capable of being marked in some way. As Schrödinger emphasized, not only are all electrons exactly alike in their observable character-istics, there is no way one electron could be marked (by being painted red, say) so that it could be distinguished from the oth-ers. One must let go of the preconception that one could pick out a certain electron (say, Ben) either by noticing some charac-teristic feature or by marking it, so that one could follow its subsequent career.

In other respects, though, the word "particle" is apt: whenever one detects *an* electron (one couldn't detect *this* or *that* electron), it always has a certain discrete signature of charge, mass, spin. It may be better to call them "quanta," to register the difference between the classical, distinguishable particle and a quantum be-ing like an electron. No human language has an adequate word for their condition. "Lack of individuality" misleadingly implies that electrons *should* have individualities, but are defective. In-stead, let me propose a new word that expresses this condition in a positive way: *identicality* means that the members of a spe-cies have identity only as instances of that species, without any features that distinguish one individual from another. Identi-cality is perfect "instantiability," the complete accord of each electron with its species in every trait. It signifies the exact nega-tion of individuality; it includes total indistinguishability and complete equality of all observable features. It has no inherent reference to time or space, and so experimentally all electrons have the same features whenever or wherever they are observed.

So strange is such a condition that some philosophers prefer to treat electrons as if they still had individualities, though hid-den from any observation by a complex web of "nonsuperve-nient relations" that are somehow embedded in physical law. This intriguing possibility would "save" the electron's individu-ality, but at the cost of introducing a plethora of otherwise in-comprehensible rules designed to hide it. If, as they concede, there is *no* known experiment that could distinguish the electron's "hidden" individuality, then I submit that in effect it has no individuality at all, to all appearances. The logical principle called "Occam's razor" tells us to prefer the simpler explanation, all else being equal; as Newton put it, "Nature does nothing in vain, and more causes are in vain when fewer suffice." For instance, when Einstein showed that a substantial ether could in no way be detected, it seemed simpler to discard it, rather than constructing complex rules to allow the ether to exist but always remain unobserved. Accordingly, I will treat electrons as if they simply had no individuality, neither "primitive thisness" nor the discernible differentiations that Leibniz expected would set one individual apart from all the rest. However, the alternate possibility of hidden individuality remains logically possible; I leave the reader to reflect further on how to judge which view is ultimately right or whether any judgment is possible.

When we start to observe these strange beings, their behavior is crucially determined by their identicality. Think of the distinguishability of each mass point that Boltzmann thought was the "first fundamental assumption" of mechanics. If electrons cannot be distinguished, how can one apply Newton's Laws to them without the possibility of confusion? Lacking any labels, one could not tell if one were really following the same electron or if it had been suddenly replaced by another. If that is so, then an electron does not have a trajectory or history in the normal sense, for that also requires being able to label each one and follow it through time.

The notion of a distinct trajectory, a unique history, is not merely an assumption of Newtonian mechanics. It is also fundamental to Einstein's relativistic mechanics, which tracks the "world-line" of each particle, its individual career through space and time. No less than Newton, Einstein also relies on the distinguishability of each particle from every other so that their world-lines do not become confused and merge, even though they become more tangled and intertwined than the densest spaghetti. If we relax the assumption that every particle is distinguishable, we open the door to a new realm that is confused in a radical sense, fusing together what used to be distinct and separate. In that realm, it is far more natural to speak of the *state* of electrons, which avoids the illusion that one can describe the precise motions of each one.

Consider a series of experiments that will illuminate the strange behavior of identical quanta by contrasting them with classical particles and waves. Imagine first shooting a stream of individual, indivisible classical particles (like tiny bullets) that are then diverted toward two opposite reflecting plates (figure 8.1).

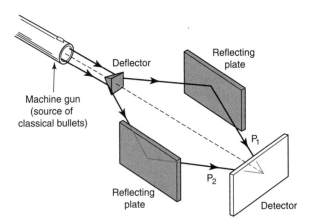

Figure 8.1
Two-path experiment with classical particles ("bullets"). P_1 is the probability of detecting a particle going along path 1 with path 2 closed; P_2 with path 1 closed; P_{12} with both paths open. Since the bullets traverse the paths independently, $P_{12} = P_1 + P_2$.

On the other side, place a detector (say a little can) to catch the reflected particles that arrive on a certain spot. By moving the detector, leaving it for an equal time at every spot, one can build up the whole pattern of the arrival of the bullets. More will arrive in some spots (especially directly along the axis of the initial beam of particles) and less in others. Consider doing this first with path 1 blocked off and path 2 open and then reversing this procedure, with path 1 open and path 2 closed. The pattern when both paths are open is simply the sum of the patterns one would get with each path closed separately, for a bullet has a distinct trajectory ($P_{12} = P_1 + P_2$). The bullets satisfy what Feynman calls "Proposition A": each travels on one and only one of the two paths. This indicates a fundamental aspect of our conception of particles, closely related to the continuity of their motion and their impenetrable hardness. Note that we assume that the bullets are equal in mass, as if they had been perfectly manufactured. Nonetheless, each is presumed distinguishable, for in reality defects of manufacturing or wear might easily render them observably different. Since the bullets are liable to such flaws, they are really not members of a species that share certain exact properties, but insofar as they approach being truly *equal,* they represent idealized Newtonian particles.

Now imagine constructing an apparatus so that we can perform a similar experiment with water waves. A train of waves hits a deflector that sends them toward two reflectors (like solid piers); the detector is a little buoy whose up-and-down motion can measure the intensity of the waves at its location. The pattern made when both paths are open (I_{12}) is quite different than when one path is closed and the other open (figure 8.2; $I_{12} \neq I_1 + I_2$). This is the hallmark of waves, as mentioned earlier. The waves coming from the two paths interfere with each other; if a crest from one path arrives at the buoy at the same

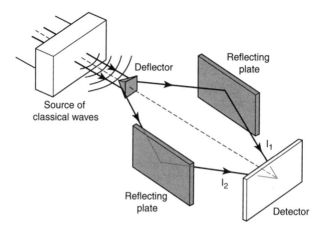

Figure 8.2
Two-path experiment with waves. I_1 is the intensity of waves observed with path 2 closed; I_2 with path 1 closed; I_{12} with both paths open. Since the waves can interfere, $I_{12} \neq I_1 + I_2$.

time as a trough from the other path, the waves cancel out ("destructive interference"). Contrariwise, two crests could arrive at the buoy, one from each path, reinforcing each other ("constructive interference"). As discussed earlier, interference is possible because waves can interpenetrate and merge since they can be *indistinguishable* when they have the same peak height (amplitude) and wave length. Yet it also follows that, since these quantities are continuously variable, waves do not fall into distinct, finite species, for they can come in infinite gradations of amplitude and wavelength.

Now imagine an apparatus that will do a similar experiment with totally identical neutrons, which are chosen to avoid the strong electrical repulsion of negatively charged electrons. Inside an evacuated vessel, a beam of neutrons (from a nuclear reactor, say) is diverted toward two deflectors (which are silicon crystals); the detector measures the number of neutrons that arrive on the other side. The two deflectors are separated by several centimeters, so that the two paths are quite distinct for the neutrons, whose beams are effectively no bigger than a small postage stamp. What should one expect? The neutrons are *equal*, like bullets, but they are *indistinguishable*, like waves. So the neutron pattern should be a synthesis of the patterns for bullets and waves, in this way: to preserve the equality of the neutrons, the detector registers only the arrival of whole, unbroken neutrons (never fractions of a neutron). However, the neutrons are indistinguishable, so it should not be possible to divide them into two exclusive classes (those that came through the upper versus the lower path), because that would require them to be distinguishable. Thus they violate Feynman's "Proposition A," since they do not travel exclusively on one path and not the other. In contrast to the bullets, the neutron pattern with both paths open should *not* be exactly the sum of the patterns of each separate slit ($P_{12} \neq P_1 + P_2$); thus the neutrons behave more like the waves and interfere with each other (figure 8.3).

Indeed, experiments confirm this pattern of interfering but integral neutrons, usually called the "wave-particle duality." Rather than trying to think of a neutron as somehow both wave and particle, it is more illuminating to see that both these aspects reflect the neutrons' identicality. To test this, let us alter the experiment slightly. Each neutron has a spin and also a magnetic moment that gauges how it responds to magnetic fields, since it has a charge distribution, though overall it is electrically neutral. It is possible to use a source that produces neutrons that are polarized, meaning that their spins are all aligned pointing up (say). If such a source is used, the same result is obtained because the neutrons are still indistinguishable ($P_{12} \neq P_1 + P_2$). But if one introduces a special magnet that will flip the spin along the lower path (figure 8.4), suddenly the neutrons

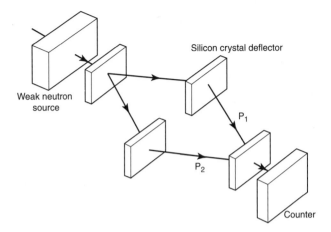

Figure 8.3
Two-path experiment with neutrons. P_1 is the probability of detecting a neutron with path 2 closed; P_2 with path 1 closed; P'_{12} with both paths open. Since the neutrons cannot be distinguished, $P'_{12} \neq P_1 + P_2$, showing a wavelike interference.

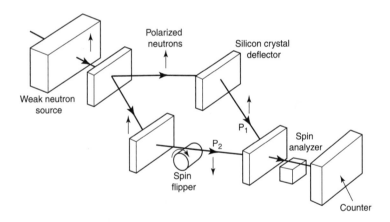

Figure 8.4
Two-path experiment with polarized neutrons (spin up, as shown). If the spin-flip magnet is off, the indistinguishable neutrons interfere ($P'_{12} \neq P_1 + P_2$). If the spin-flip magnet is on, the neutrons on path 2 are distinguishable from those on path 1 and there is no interference ($P'_{12} = P_1 + P_2$).

from the two paths are quite distinguishable and the pattern of interference disappears ($P_{12} = P_1 + P_2$); in this case, they act just like bullets!

This is even stranger when one realizes that neutrons are not elementary, for they are composed of three quarks. The identicality of each neutron means that the quarks of each species are identical and combine to form identical collective states that we call "a neutron." These characteristic quantum interference effects also appear in other, larger systems. Similar experiments have been performed with atoms, small molecules, and even "buckyballs," sixty carbon atoms arranged in a soccer-ball configuration.

Returning to the neutron experiments, it is important to emphasize that the source is so weak that there is only one neutron in the apparatus at any given time. That is, the issue is not primarily the indistinguishability of any two neutrons, but even more the continued identity of any neutron *with itself* over time and through space. At first glance, this would not seem to be a problem, but, as Locke earlier realized, it is. Just as with the ship of Theseus, a bullet tends to become worn and dented, and even a neutron might conceivably lose mass in collisions with other neutrons, though that has never been observed to occur. Yet surely its identity depends on its mass or other observable properties; only if those properties are slightly variable could we ever distinguish one from another, thus differentiating the individuals within their presumed species. Contrariwise, if neutrons truly form a unique species, their identity cannot vary over space and time or suffer any experimental alteration. Thus even a single neutron maintains its identicality, refusing to submit to "Proposition A" by not showing a localizable, unique path.

This concept has some subtlety. Thinking of Millikan's experiment measuring the charge on oil drops, it *is* possible to distinguish an electron on drop A (in one room, say) from one on drop B (in another apparatus, in a different building or even a different galaxy). Yet this really distinguishes not the electrons but only the drops each is on, which are sufficiently macroscopic to be distinct, at least by virtue of the room they are in. Thus, though one can distinguish electrons by such equivocal means, by reference to larger structures containing them, there *are* situations in which the true and intrinsic identicality of the electrons cannot be avoided, as when one brings electrons close enough so that they are not identified with their separate drops.

This precise sense of identicality needs to be kept in mind as one looks at photographs taken in a cloud chamber that seem to indicate the trajectories of distinct particles (figure 8.5). The

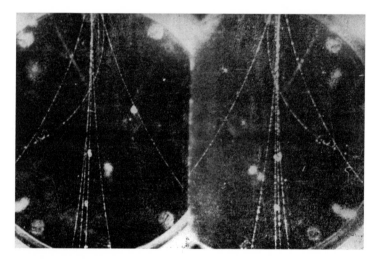

Figure 8.5
Cloud chamber tracks of elementary particles, photographed by Carl Anderson in 1936. This is a stereoscopic view of a shower of electrons and positrons (positively charged antielectrons), whose tracks curve in opposite directions in a magnetic field.

photograph does not really show each electron or positron, but only the wake it has left in passing. Each charged particles ionizes the air and leaves a tiny contrail around which the visible cloud forms, akin to the formation of large-scale contrails in the wake of supersonic airplanes. What we see reflects atomic events magnified through a cascade of processes capable of rendering an image so huge that even gross human perception can grasp it. There is even more subtlety in scanning tunneling microscope (STM) images of a single atom (figure 8.6). Here, the image that we see represents the electric current flowing between a needle and the scanned surface. To make these measurements visually intelligible, they are translated arbitrarily into a computer reconstruction, like a contour map whose "false color" helps the eye grasp the information. As faithful as it is, this is not a really naïve "picture" of an atom, compared with, say, a visual image of a snowflake gained by direct reception of scattered visible light.

The very process by which the STM works involves the passage of innumerable identical electrons from the surface to the scanning needle, so delicately that their current measures the precise contour of a single atom. More startling still, an atom can be picked up and moved by a "magic wrist," a robotic manipulator attached to a STM. Even so, the atom has not been marked except by its location in a vastly magnified image, identified only in relation to macroscopic objects. If we were to turn the needle away, we could never retrieve that atom again.

This lack of individuality also haunts the amazing images of a single atom, ingeniously penned into a tiny trap, kept far apart from other atoms. Here one can actually *see* a single atom, not just a computer-generated image (figure 8.7). Its intermittent twinkling is a direct manifestation of quantum jumps as the

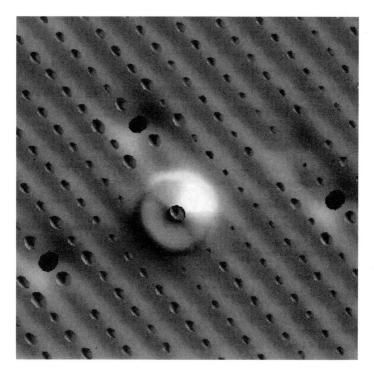

Figure 8.6
An image of a single xenon atom on a crystal of nickel, taken by Don Eigler (IBM) using a scanning tunneling microscope. The computer reconstruction has left a hole in the center of the xenon atom, through which one can see the nickel atom directly underneath.

atom absorbs light, becomes excited, and radiates away its excess energy. Such an atom can remain trapped for months, becoming so familiar to its handlers that they give it a name like "Astrid," for its starry glittering. Yet it remains wild and without individuality, however long they try to tame it. If you let "Astrid" go free, there is no way to trap it again. The name is human, the atom nameless.

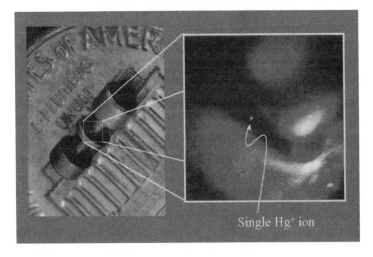

Single Hg⁺ ion

Figure 8.7
A trap designed to isolate single atoms, shown actual size against a penny. On the right, a video image made by David Wineland (NIST) shows a small dot inside the trap, which is the direct light from a single isolated mercury ion.

This cannot be evaded by making the trap ever smaller. The smaller the box, the stronger the forces one will need. The more we try to confine it, the faster and more irregularly the atom moves, as if it were evading being pinned down. This anthropomorphic way of speaking is not correct—an atom has no volition or sensation, of course—but it does give a vivid and intuitive form to what is usually called the Heisenberg uncertainty principle: the more one localizes an atom in space, the more its momentum becomes uncertain. In fact, if one pursues the project of trying to localize an atom past a certain point, the forces used to confine it ever more tightly push it to speeds approaching that of light. At such speeds, the theory of relativity enters in (as I will discuss in chapter 10), for one has to use so much energy that pairs of electrons and antielectrons start to spring up inside the box, generated by the very field used to try to keep them out. Even if one just uses ordinary matter to build the box, that matter is built from atoms identical with the prisoner; as the box becomes ever smaller it becomes less possible to assert that the prisoner could not have exchanged places with one of the guards.

It is exactly this possibility of exchange that prevents one from treating an atom as separate from its brethren. Like identical twins, they are always fooling us by switching places. Their intense intimacy cannot be breached by others, however perceptive. This metaphor is not exact, since human twins are not as identical as atoms, yet it may convey something helpful. Confusion of identity is manifest as weird quantum effects, the "spooky" correlations that Einstein found so outrageous, as if they were the mischievous pranks played by twins. His favorite example involves an excited atom that emits a pair of particles on directly opposite courses, traveling far away from each other and from their parent atom. Because of the conservation of total spin, each of the emitted particles is precisely correlated to its distant twin, so that if one detects that the spin of one twin is up, then one instantaneously knows that the spin of the other twin is down, without having to measure it.

This bizarre circumstance is the celebrated paradox that Einstein, Boris Podolsky, and Nathan Rosen published in 1935. Though the correlation of the twins seems to violate the theory of relativity, it does not, because no information can be transmitted using the twins. Their conspiracy is wholly playful, for it contains no message beyond their perfect coordination; when one twin is up, the other is always down and vice versa, but the sequence of spins of either twin is random. This particularly bothered Einstein, who insisted that "God does not play dice."

Even at his boldest, he remained faithful to Leibniz's vision of preestablished harmony ruling the world, allowing nothing random or indeterminate. He insisted that physics must give "a complete description of the individual system," ultimately determining the path of every particle without recourse to randomness. Of course, this would require the identification of each particle as a unique individual. As the next chapter will explore, Einstein's vision may have failed him just on the issue of individuality.

Beyond their importance as philosophical battlegrounds, these uncanny effects are not mere theoretical oddities but can be experimentally tested. Through ingenious techniques, Einstein's paradox has been reenacted in the laboratory with results that completely confirm the predictions of quantum theory, however strange they may be. Such experiments also confirm that theory through testing the inequalities derived by John S. Bell. Yet the true dimensions of quantum identity have superlative practical importance going far beyond subtle paradox. The structure of each chemical element, the periodic table built from those elements, and the chemical bond itself all crucially depend on the correlations due to identicality, as will shortly emerge.

Einstein's objections stemmed from his belief that, ultimately, the career of each individual particle is perfectly determined by the laws of physics. Identicality requires that we go beyond the way Newton described identifiable particles, each having an observable position and velocity governed by a mathematical equation. This leads to a profoundly new mathematical picture of quantum theory, first expressed by Born in 1928, which I shall elaborate and extend. In order to hide individuality, the world has two levels: an outer level of positive numbers, which represent the probability of observing an electron somewhere in space-time, and an inner level not accessible to observation but that guides the observable probabilities. On the inner level, there are *amplitudes,* complex numbers (that is, involving the square roots of negative numbers), which cannot be observed, for our measuring devices register only real numbers. What is more, Born noted that these amplitudes follow strict equations (such as Schrödinger's equation, or Dirac's) that determine how they unfold over time with *no* uncertainty.

However, the probabilities that we observe turn out to be the absolute squares of these amplitudes, and these probabilities *do not* unfold in a determined way: for instance, we cannot determine which path the neutron took. Even if we know the neutron's *amplitude* perfectly, the probability of its being observed at one time and place does not determine, absolutely, the probability of its being observed later, at another time and place, for that would mean identifying the neutron. So there is no master equation for predicting the appearances of the neutron with certainty, even though there *is* an equation enabling us to predict the *probabilities* that we can observe the neutron here or there.

Since the probability of observation is a square of the amplitude (and hence always positive), there is the mathematical possibility of the amplitude being either positive or negative, to take the simplest case. This mathematical possibility is related to an important further ramification of identicality: identical quanta come in two different large groups, according to the symmetries of their amplitudes upon the exchange of like particles. Those with symmetric amplitudes (which are unchanged upon interchange of the quanta) are called *bosons* (for "Bose-Einstein statistics"); they include light (photons) and all quanta whose spin is an integral multiple of $h/2\pi$. Planck encountered this symmetric indistinguishability when he introduced his novel counting. The other group of quanta have antisymmetric amplitudes (which are reversed in sign when the quanta are interchanged) and are called *fermions* (for "Fermi-Dirac statistics"); their spins are half-integer multiples of $h/2\pi$ and include the electron, the proton, the quarks, etc. (If the quanta are confined to a two-dimensional surface, there are other possibilities, called "anyons.")

Because of the symmetries of their amplitudes, the bosons prefer to share the same state, as much as possible. This turns out to be crucial to the functioning of lasers, in which coherent light emission depends on light quanta preferring to crowd together in one state. On the contrary, fermions can't stand to share the same state (a situation that is called the "Pauli exclusion principle"). A simple example may illuminate this. Two ordinary coins, when flipped, would yield the four possibilities head–head, head–tail, tail–head, tail–tail, each of which we would statistically expect to observe with equal frequency, $1/4$. Thus the odds of getting a head and a tail in either order should be $1/4 + 1/4 = 1/2$, compared to $1/4$ for two heads or for two tails. Note that this requires the two coins to be distinguishable. If, on the other hand, we had two "quantum coins," both identical bosons, we would expect only three possibilities—head–head, one head and one tail, and tail–tail—because now we cannot tell head-tail from tail-head. We would expect each of these three to occur with equal probability, namely, $1/3$. Thus in this case the odds of getting one head and one tail go down to $1/3$, the same as the odds of getting two heads or two tails in this case. So the probability of two bosons being in the *same* state ($1/3$) is higher than that for two distinguishable particles ($1/4$), which is crucial to the operation of the laser. In contrast, two "quantum coins" that are fermions could never show both heads or both tails, for that would require their sharing the same state, which they cannot do. This is the crucial fact in the building of the periodic table of elements.

Consider, for example, the elements past hydrogen, which has one proton in the nucleus and one orbital electron. As one tries to add another electron, the exclusion principle keeps that second electron from sharing the state of the first. This can be done if the two electrons have opposite spins but occupy the same general zone around the nucleus, a "closed shell" in chemical terms. This second element (helium) accordingly is inert chemically, both electrons remaining quite close to the nucleus, not readily available for chemical interactions with other atoms. In contrast, if one adds a third electron, it must keep away from the first two (since there are only two spin possibilities for electrons, both now taken) and is thus available to interact with other atoms, leading to a highly reactive element (lithium). Similarly, as one adds electrons, one successively generates each new element and its chemical properties. In this way, electrons may form more or less closed shells in atoms, giving each element its characteristic structure and also giving the recurrent patterns of the periodic table. Thus the mathematical possibilities opened up by the two great modes of symmetry in identity (bosons and fermions) have left their signatures everywhere.

Mathematics is only a *representation* of physical reality, but as such it gives an insight into the reality it is representing. In the process, it dispels certain widespread misunderstandings. For instance, quantum theory is not finally about uncertainty and indeterminism. On the inner level, quantum theory is as certain and determined as the theories of Newton and Maxwell. The apparent uncertainty emerges only when we try to observe outwardly these inwardly determined things. After all, though we are made of identical electrons (as well as the quarks constituting our protons and neutrons), we are extremely large, meaning that we are composed of a huge number of these identical components, which can enter into extremely complex states. It is those complex states that "are" us, in the sense of our apparent overall individuality, and they characterize all our attempts to observe the world. When we turn to observe a single atom, it is with huge devices, compared to the electron, attuned to our enormous size and crude sensibilities. Yet our minds can enter a realm of certainty through the equations of the theory.

Still, such a double view of the world is disturbing. For Einstein, it abandoned the central project of physics: the complete understanding of observable data through mathematical theory. He would not accept Born's view that what is observable is not certain, and what is certain is not observable. To be sure, Einstein recognized the power and validity of quantum theory, though he did not regard it as complete. His concept of understanding demanded going beyond these apparent limits to knowledge. He thought that each electron really has its own individuality, which *we* are confused about, but which God knows perfectly well. Even as great a man as Einstein finally had his limits, which were part of his strength, defining the credo he lived by and the ideas he could not endure. Yet something always lies beyond. The electron's identicality runs so deep that *no* observer, not even God, can distinguish one from another. It is no wonder that Einstein resisted this idea, for it is genuinely strange. However, as Francis Bacon observed, "there is no excellent beauty that hath not some strangeness in the proportion." Viewed a different way, the strangeness of identicality might lead us to see its beauty.

According to an old tale, there were 36 universes created before the present one. For mysterious reasons, they did not survive. What happened to them? Were they unworkable or just ugly? Were they destroyed by a disappointed creator or did they self-destruct? Einstein spent his life wondering what were God's thoughts as He contemplated cosmos number 37. Was it a matter of adjusting certain basic physical constants? Did the angels suggest different physical laws? Did He have any choice at all? Though Einstein missed the possibility, one may well wonder: why did God choose identicality?

9

The Silence of the Sirens

The idea of identicality is so strange that one must look far afield to find a way to grasp it intuitively. After all, it is hard to put aside all the preconceptions and familiar assumptions that go along with our everyday experiences and our assumptions about individuality. In part, Einstein was following a prejudice deeply ingrained in human nature. Yet there were other thinkers in Einstein's sphere who might have changed his mind. In the ideas of Einstein's favorite philosophers, there are tantalizing indications that open new perspectives toward the disturbing ideas of quantum theory. Among his acquaintances, Franz Kafka wrote stories that turn on strange displacements of identity and, in so doing, may give a helpful parallel to the quantum realm.

As a young man, Einstein joined with two other friends to form what they humorously named the "Olympian Academy," making fun of pompous academic institutions. They would meet in a coffee house to discuss books that excited them and these meetings were important to Einstein's development; their discussions were animated and wide-ranging, liberated from the conventional boundaries of disciplines and academic shibboleths. They would read philosophy, physics, and literature; Einstein always remembered those meetings with delight and appreciation. Later in life, he found that good conversations are rare, even in famous academies. Certain threads in Einstein's early reading could have shown a different view of individuality to him, which we might follow even if he did not.

The writings of Immanuel Kant had long fascinated Einstein; he began to read him as a teenager and the Olympian Academy turned to this philosopher as if he spoke immediately to their concerns, though he lived a century before. Kant is often represented as a conservative and methodical Prussian, famously disciplined and punctual, a champion of Newtonian physics and an indefatigable system-builder. Yet he was a radical; he would drink wine to celebrate Bastille Day each year and wrote eloquently about the enlightenment of man. *Dare to know* was his motto; he wanted humanity to wake from its childish and dogmatic slumbers. He appealed to Einstein because he articulated a radical view of physics, one that stemmed from Euclid and Newton, but which can be read also in the light of quantum theory.

Kant argued that physics reveals the outer appearances of the world. The inner nature of bodies, even of our own bodies, remains forever hidden: we cannot say what they are in themselves, apart from how they appear to us. This hidden side of things Kant called the *noumena,* using a Greek word signifying the "intelligible things," in contrast to the *phenomena,* which signify the "apparent things" that we can observe. Ironically, we can know nothing definite about the "intelligible things," though we can speculate at will. The phenomena, in contrast, follow physical laws and are open to our investigation; we cannot, however, discover what lies behind the phenomena. Kant compared our knowledge to an island surrounded by unfathomable waters, representing the impenetrable hiddenness of things in themselves.

The impetus for this disturbing idea came from Newton, who warned that he would not invent hypotheses about the causes of gravity or inertia, but would treat only the outward mathematical appearances measured by mass. Newton may have hoped that he or others could later return to these questions and give definitive answers; a certain melancholy in his later writings suggests that he may have decided that no one, ever, could penetrate the hiddenness of nature. Kant gave forceful expression to this deep limitation of Newton's project, making conscious what Newton only glimpsed or feared, ascribing it not to any flaw in Newton's approach but to ultimate limitations of reason itself. Indeed, Kant considered that only through acknowledging this limitation could natural science progress, liberated from the false dream of penetrating beyond appearances.

Though Kant approached these matters from a Newtonian point of view, his insights are even more provocative in light of quantum theory. Quantum theory also has two levels, which can be compared with Kant's division between the intelligible world and the sensible world. First, there is what I earlier called the outer level of observable probabilities, which can be compared to Kant's phenomena. However, in contrast with Newtonian physics, in quantum theory this outer level is *not* deterministic. On the other hand, there is the inner level of the amplitudes, unobservable but completely intelligible through the equations of quantum theory. In contrast, Kant's noumena lie in a realm completely beyond observation but available in a way to the mind, a realm in which one could believe what one wanted about the inner life of bodies, but could never really find out. Kant and quantum theory disagree about whether anything can be said about this hidden intelligible realm. They agree on a simpler but still important thought: there is something deeply hidden behind things.

Looking out from the island of knowledge, Kant glimpsed new possibilities of identity. Though he admired much about Leibniz's philosophy, he criticized Leibniz's principle that no two individuals could be identical. Leibniz had treated the appearance of individuals as if it presented the whole of their nature, summed up in the "complete notion" of all they ever did or will do. Kant argued that this does not speak to each thing, in itself, apart from all others. Because of the unknowableness of that hidden realm, Kant noted that Leibniz's principle cannot simply be considered a law of nature, but only a teaching about how things *appear* to us. Kant also argued that one could *imagine* two indistinguishable drops of water, even if one never had happened to see them, and that alone would show that Leibniz's principle is not absolutely necessary. This argument remains inconclusive because, in some imaginary world, a weird natural law might cause one to perceive two apparently identical replicas instead of a single drop. Leibniz also had offered purely logical arguments but felt the need to check whether or not there were identical leaves or water drops in *our* universe. Of course, Kant had no grounds to suspect that there were things like electrons that are identical even as phenomena and he certainly did not anticipate quantum theory. But by noting the divided quality of the world and the unknown depths of identity, he pointed in the direction that quantum theory would take.

As much as he admired and pondered Kant's theories, Einstein's favorite philosopher was Benedict Spinoza, a holy heretic who had been anathematized by Jews and Christians and whose apartness and intense purity touched Einstein deeply. Einstein was moved by Spinoza's vision that everything is a mode of the one, infinite God, whether manifest as matter or as mind: everything in the universe is a single, unified substance in which there is no accident or randomness. Einstein took Spinoza's statements about the perfect determination of this god-universe as confirmation of his conviction that physics must keep faith with determinism.

However, his favorite philosopher may have opened up pathways that Einstein either did not notice or hesitated to enter upon. Spinoza did assert that "things could have been produced by God in no other way, and in no other order than they have been produced," and offered a proof modeled on geometrical arguments. But Einstein did not realize that Spinoza also divided the simple concept of nature into two aspects: apparent nature ("nature natured," in Latin *natura naturata*) and the depths of nature ("nature naturing," *natura naturans*). Drawing on this medieval terminology, Spinoza indicates that "the ideas of the affections of the human body, insofar as they are related only to the human mind, are not clear and distinct, but confused." Though we humans (like everything in the universe) are a part of God, we are *only* a part, and hence our view of things is limited and confused: for us, "all particular things are contingent and corruptible." Our observations are limited to the sphere of apparent nature, "nature natured," as it emerges under our groping investigation. Our knowledge is merely probable, not certain; this is similar to the probabilistic outer level of the world, according to quantum theory. On the other hand, God or Nature is perfectly determined, in itself, on the deep level of "nature naturing," and on that level perfect determination must hold sway. This is the inner level of quantum theory, the level of the mathematical equations and the amplitudes they determine so exactly.

Viewed in this way, Spinoza seems to anticipate the two levels of quantum theory uncannily. Indeed, Kant's division of the world was deeply influenced by Spinoza, whose version may be even closer to the quantum view. For Kant, nothing at all can be said about the noumena, the intelligible things; for Spinoza, there must be a perfect coherence to the depths of "nature naturing," even though it is obscured by the partiality of our view of "nature natured." Spinoza might have been pleased that the determinism of the quantum equations is hidden from direct observation under a screen of probabilities. But the deepest and most exciting part of Spinoza's vision is his radical view of individuality. He acknowledges that, on the level of appearances, there seem to be individuals, with their own particular nature: Peter is not just an instance of the idea of man but of a more particular idea of Peter. However, Spinoza's deepest thought is that all bodies must be understood as modes of the one, unique substance he calls god or nature (*deus sive natura*): "the whole of nature is one individual, whose parts, that is, all bodies, vary in infinite ways, without any change of the whole individual." This stunning idea has powerful consequences: what we call separate bodies cannot really be individuals, for they are aspects of the one true individual, call it god or nature. When we talk about this or that body, what we really mean is their motion or rest, speed or slowness, and not their substance, *which is one.* The radical individuality that Leibniz claimed for every body is, for Spinoza, an illusion, a mere appearance. This is very close to identicality, and Spinoza formulates it with boldness and rigor.

Perhaps the Olympian Academy broke up too soon. It should have kept discussing Spinoza; Einstein's favorite philosopher could have helped him more than he knew. Indeed, Einstein continued to think about Spinoza to the end of his life. In 1951, just after the death of his beloved sister, Einstein wrote that "we bear many afflictions unflinchingly, but Spinoza's precarious God has made our task more difficult than our forefathers suspected." Einstein's meaning is tantalizingly unclear, but Spinoza seems to trouble him, as if simple resignation through complete determinism is not what that "precarious God" prescribes. Perhaps it was his own convictions that Einstein was finding precarious.

Einstein's contemporaries did not cease trying to reconcile him to quantum theory; Niels Bohr recalled that they spent much time in 1937 humorously arguing about whose side Spinoza would have taken had he joined their quantum debates. Einstein knew someone else who described a strange realm of identity: Franz Kafka. In the years 1910 to 1912, Einstein and Kafka both lived in Prague. They met, shared friends, and frequented the same salon. There are reflections of their encounters in the writings of those who knew them, including their mutual friend Max Brod, but it seems that they were too different to be close, though many strands drew them together. Both of them struggled with what Einstein called the "merely-personal": the entrapping web of personality, the trivialities and pettiness of the world, the horror looming over Europe. Einstein sought escape in science and music, solace that Kafka almost envied, had he not deeper reservations about all human striving. He remarked that "we live in straight lines, yet every man is in fact a labyrinth." This inner labyrinth is constantly rising up to trouble the straight lines within which we try to capture reality.

Kafka's world is close to the strangeness of quantum theory. His characters grope in an uncertain outward realm of appearances while sensing dimly the hidden inwardness of the Law they seek. In *The Trial,* Joseph K. is accused of a mysterious, unnamed crime and learns that there is "ostensible acquittal" and "indefinite postponement" as well as the endlessly elusive "definitive acquittal." The world is divided into the ostensible and the definitive; this division reflects the dark possibilities of human nature. We are so predatory that the Creator hides from us: "God must remain hidden in darkness. . . . The interrelation of objective and personal values doesn't function any more." Accordingly, the search for the purely objective has something delusive about it. As Kafka phrased it, we seek heaven in a drop of water, but that drop yields "an image which is distorted and blurred by our slightest movement." Since the particular quality of subjective observation decisively affects the outcome, observation affects what is observed, as in quantum theory. In his story "The Silence of the Sirens," Kafka reverses many of the details that Homer recounts in order to show the power of the subjective. Where Homer's Sirens were dangerously powerful, Kafka's Sirens are lost in Ulysses' gaze. "They no longer had any desire to allure; all that they wanted was to hold as long as they could the radiance that fell from Ulysses' great eyes." Kafka's Ulysses may be a faker, even a childish deceiver, but his unfathomable inner state has immense power to reshape appearances.

Compared with Einstein's optimistic vision of the transcendent Law, Kafka's view is dark. The mysterious law courts in *The Trial* are filthy, the lawyers lecherous. Our world is a penal colony, justice a horrifyingly cruel, broken-down machine. We are drawn to the Law but stand at its gate filled with fear. It is no wonder that Einstein returned *The Trial* to a friend who loaned it to him, confessing that he could not read the book. These feelings of fear transcend the "merely-personal," but in a way that is different from Einstein's mathematical expressions. For Kafka, such everyday emotions as pride or anxiety are connected inextricably with the operations that draw us to the Law; in *The Trial,* the mysterious law courts occupy the recesses of ordinary buildings. Kafka said that "men can achieve greatness only by surmounting their own littleness," but those limitations always surround us as the prison bars of our human situation. It is our awareness of them and the attendant reverence for what is higher that may save us. For Kafka, as for Dostoyevsky, this generally takes the form of guilt and redemptive suffering: we have sinned and we must die. "The iron fist of technology destroys all protecting walls. . . . We are driven towards truth like criminals to the seat of judgment. . . . We are disturbers of order and of peace. That is our original sin. We set ourselves above nature. We are not content to die and to survive merely as members of a species."

For Kafka, each act that takes us out of the anonymous wholeness of nature and individualizes us is part of our "original sin." Here he seems to mean something like Arthur Schopenhauer's sense of the fatality of human individuation, of removing ourselves from the original oneness of nature and assuming willful states of differentiated individuality. For Schopenhauer and also Kafka, music represented one of the ways in which man could overcome this individuation, reaching past individuality to a deeper, wordless unity.

In Kafka's celebrated story "The Metamorphosis," a traveling salesman wakes to find himself transformed into a loathsome insect. Gregor Samsa's voice and identity remain intact at first, though gradually yielding to his new form. To his horror, he is ravenous to eat the repulsive garbage that his insect body craves. Gregor suffers a deep transformation of his original identity as a rather ordinary and selfish man. As an insect, he maintains his human memories but also reaches toward a kind of nobility and selflessness he never knew as a man. Out of self-disgust and heartbroken love for his family, Gregor starves himself. Approaching death, he begins to feel an intense connection with music, as if it contained the secret of his transformed identity. At the climactic moment, he crawls into the next room to hear his sister's violin playing, unable to control his need to listen to her music even though his appearance disgusts his family. Gregor moves toward music "as if it were the unknown food he craved." Kafka often lamented his own unresponsiveness to music as a measure of his grave condition, for it showed how grievously he lacked a vital connection with the primordial unity of things.

As Kafka represents it, the fundamental and original state of the world is an unbroken wholeness of beings that are not distinct. They are numerous but still identical, like the three visitors that came to Abraham in his tent or the defendants sitting in the shadowy antechambers of the law courts or the silent sirens. The protagonists in Kafka's later works have less and less distinct individuality. Joseph K. has no surname and the land-surveyor K. in *The Castle* has no name at all, only an initial, as if it were an algebraic variable generalized from Kafka's own name. The Castle itself is perhaps the best example of the realm of the identical; when he calls it on the telephone, the land-surveyor hears an indistinct mélange of laughing voices. K. cannot gain entrance to it; he cannot even stay awake during a climactic conversation in which an official attempts to explain the mysteries of the Castle to him. As he tried to enter the realm of the identical, sleep overcame his spirit.

Entrance to the realm of the Law means leaving behind individuality, whether the "merely-personal" littleness of the seeker or the individual distinctions that pervade the familiar world of experience. Einstein joins Kafka in saying that "reality is unreal," that the realm of the Law is necessarily strange for us. During the first part of his life, Einstein felt his journey toward the Law to be a happy and liberating experience. It was only toward the end that the heavy weight of history and his growing sense of responsibility dimmed his happiness. At that point, he remarked that he wished that he had been a plumber instead of a professional physicist. Kafka loved carpentry as an avocation and said that "intellectual labor tears a man out of human society. A craft . . . leads him towards men."

Both men lived lives of the greatest singularity and yet wanted to submerge themselves in human anonymity. This is less surprising in view of their devotion to a realm of law beyond individuation. Both of them confronted their personal failures as lovers and family men. Though Max Born considered Einstein his dearest friend, he noted that "for all his kindness, sociability, and love of humanity, he was nevertheless totally detached from his environment and the human beings included in it." Einstein recognized that he had "twice failed rather shamefully" to sustain a marriage and Kafka excoriated himself for his failure to marry at all. Doubtless this touches on personal matters beyond the scope of their work. Yet there may also be a deeper connection between their personal experiences and their quests to go beyond the "merely-personal" into a realm beyond ordinary individuality. Kafka remarked that "ideas—like everything else in the world which has a super-personal value—only live by personal sacrifices." The call to rise above the "merely-personal" finally raises the deepest personal questions.

10

Beyond Being

The radical loss of individuality discloses a new depth behind things. It may lead to further surprises, entering a realm so distant from our preconceptions. Though it is reason that guides our halting intuition, the ordeal of experiment must test and purge our expectations. If even one electron proves to be truly distinguishable from the others, everything must be reconsidered. The touchstone of experience must be applied unsparingly and without end. Yet, as Bacon realized, beauty and strangeness go together. What is strange is foreign or alien, *étrange*. To macroscopic beings composed of vast numbers of atoms, large enough to consider themselves unique, the realm of identicality will always seem foreign and hence both weird and intriguing. As the Greeks knew, the stranger may be a god, come from beyond our horizon to test our discernment and our hospitality. As Hamlet teased his skeptical friend, "There are more things in heaven and earth, Horatio, / Than are dreamt of in your philosophy." However strange these things may be, we too may find ourselves to be strangers in the world, or to new truth, "and therefore as a stranger give it welcome."

So far, the identicality of each known species of matter has withstood arduous testing. For instance, the Pauli exclusion principle (which depends on the strict indistinguishability of electrons) has been tested and found to hold within one part in 10^{34} (1 followed by 34 zeros). Even if someday such tests discover some measurable variation in the properties of electrons, this incredible near-equality will remain of capital importance, a touchstone for further theory. As it stands now, there is no experimental evidence against identicality to a degree that goes far beyond any human artifice.

Though the preceding chapters have indicated some of the strange consequences of this lack of individuality, there remain others, some probably still undiscovered. Already identicality has implied the blending of particle and wave, previously so antithetical, into behavior that has no parallel in classical physics. This confusion has broad implications even for the basic concept of number. If it is true that we cannot distinguish one electron from another, then it must follow that we cannot count them in the ordinary way. The reason for this is that when we count, we must rely on differences between objects to keep from counting the same object over and over, under the mistaken impression that it has not yet been included. In principle, we mark and set aside each object as it is counted, in order to ensure that no double counting takes place. Of course, this is not possible for electrons.

Yet this indistinguishability does not mean that there is not a definite number of electrons present. The law of conservation of charge is still fixed in the quantum theory of electrons, and it requires that the total number of electrons in any closed system remains constant. In fact, there are experimental techniques for determining this number, such as those Millikan used, which, he proudly announced, could count the charges on his drops "no less surely than you could count your fingers and your toes." However, he was not counting them so literally; rather, he determined the total charge on the drop and divided it by the unit of charge that he had determined over many trials. The result of this calculation was clearly the number of electrons on the drop, though they never were directly counted, nor could they have been.

To put it in the language of arithmetic: Millikan found the *cardinal* number of charges on the drop, which answers the question: how many? He could not have found the *ordinal* number, which counts objects by placing them in order of succession: first, second, third, . . . Yet in ordinary arithmetic both these sorts of numbers coincide, for we usually count off objects one by one, ordinally, so that when we stop, we have the cardinal number of how many objects are present: one, two, three. On the face of it, having a cardinal number without an ordinal number seems very strange. It is even stranger than Lewis Carroll's Cheshire cat, which could disappear but still leave its smile, for in the case of electrons, the smile (cardinal number) exists without *ever* having had a cat (ordinal number). If this were proposed as a bizarre property of some new mathematical system of "non-ordinal numbers," one would think it far from common experience. Yet the very electrons that compose us obey this strange arithmetic.

In a direct and powerful way, identicality indicates a new kind of mathematics that rethinks the most basic truths. For the Greeks, the counting numbers were the origin and touchstone of all mathematics, even of all knowledge. It would take another book to tell the story of the very different quantities we now call "real numbers." These include, along with integers, the irrational numbers that so disturbed and amazed the Greeks: "algebraic numbers" (such as the square root of 2 or, going beyond Greek geometry, the solutions of any finite algebraic equation) and "transcendental numbers" (those not the solution of any finite algebraic equation, such as pi). Euclid, the master synthesizer of Greek mathematics, would never have recognized these as worthy of the name of numbers; he would have called them "magnitudes," like the lengths of geometric lines. He would have been further astonished by the introduction of "imaginary numbers" (like the square root of -1), "ideal numbers" (higher roots of 1), and quaternions or "Cayley numbers" (generalizations of imaginary numbers that do not commute: **a** × **b** is not equal to **b** × **a**, as is also true of matrix multiplication). But if he had heard of numbers one cannot count, he might just have laughed.

At the least, this strange arithmetic indicates that identicality emerges from a realm quite other than common numbers. Indeed, the very idea of physical objects so inhumanly identical suggests a transcendent domain of mathematical forms that stands apart from ordinary experience. As the new quantum theory of 1926 came into being, it called for a mathematics that would do it justice and found it in the equations of particles and fields known to Maxwell. By a strange and beautiful coincidence, the mathematics needed had already been developed in the preceding century in the context of pure speculation, with no thought that it would ever be used in physics. The mathematics of quantum amplitudes rests on noncommuting "numbers"; **a** × **b** not equaling **b** × **a** reflects the experimental reality that the order in which measurements are performed is crucial and not reversible. This is the core of the uncertainty principle: measurement irreversibly affects what is observed. Such unconventional mathematics caused some consternation even to the founders of the new theory, who were not all versed in these arcana. When informed that what he had just discovered was really matrix mechanics, Heisenberg wailed: "But I don't even know what a matrix is!" Schrödinger tried his best to make his wave mechanics follow the pattern of ordinary wave equations for water or sound. With a certain horror, he was forced to acknowledge that his "wave" did not inhabit the three-dimensional space of our experience but a "configuration space," a theoretical construct having three dimensions *for each particle,* rather than a normal three-dimensional space shared by all the particles.

There was already some precedent for these extravagant mathematical structures in the development of statistical physics. Boltzmann and Gibbs had pioneered the use of "phase space" to describe gases or other systems with large numbers of particles when one is concerned with the behavior of the whole ensemble, not individual atoms. There had to be six dimensions of phase space for each particle in the gas (three for the position of each particle and three for its momentum), so one would need 6×10^{25} dimensions to describe a cubic meter of air. Yet these dimensions were a *representation* of the gas, not to be mistaken for the gas itself. A map of Santa Fe should not be confused with Santa Fe itself. In a similar way, elaborate economic models use different financial indices as "dimensions" to describe the performance of the economy in a multidimensional economic "space." These "dimensions" could be as diverse as the price of gold, the prime interest rate, the gross national product. Still, no one believes that the economy really is a spatial object; when the Dow-Jones Index goes "up," nothing has *physically* risen. Somehow, it is harder to maintain this detachment when physical objects are under consideration. Schrödinger's impulse was natural: if electrons really were described by his wave equation, their waves ought to be physically present in space. His attempt to explain quantum phenomena purely in terms of waves fell afoul of a fundamental divergence between the representation and the palpable reality.

To give an adequate representation of electrons, it proved necessary to invoke not just a "space" of high, but finite, dimensionality (like phase space) but one that in principle had an infinite number of dimensions: Hilbert space. That is the true realm of the quantum; its reality goes beyond existence in the ordinary sense of existence-in-space-and-time. Philosophers have pointed out that universal terms (like "goodness") are neutral with respect to existence, properly speaking; universals neither exist nor do not exist, because they are not truly situated in space and time. In contrast, individuals simply *exist,* whether persons, things, or their features. Since electrons are not individuals but instances of universals, *they too are neutral with respect to existence.* Strictly speaking, then, we should not say an electron "exists," nor should we say the opposite. On the other hand, electrons are very real, but in a way that goes beyond our ordinary concepts, including that of existence, gauged for the macroscopic world of individuals.

In their internal realm, electrons enjoy "internal symmetries" like gauge invariance, which are outwardly manifest to us as the conservation of charge. One might even go so far as to say that, as the speed of light connects space and time, Planck's constant connects the inner and outer realms of the quantum. In both cases, a universal physical constant bridges two realms that had seemed incommensurable.

Here considerations of counting can extend these insights. When energies are low, the number of electrons is strictly conserved, meaning that none is created or destroyed. Einstein's celebrated equation $E = mc^2$ stipulates that one would have to provide energy E to create a body of mass m. In the case of electrons, this represents a great deal of energy, available in nuclear reactions but not in chemical ones. In contrast, quanta of light, being massless, can be created and destroyed constantly, even at low energy. Every object is continually radiating and absorbing light, even in what seems to us total darkness, which is filled with infrared and radio emissions our eyes cannot perceive. Already in 1905, Einstein pointed out that light behaves as particles in the photoelectric effect. Thus, since the number of light particles (or photons) is not conserved, the theory of relativity opened the possibility that electrons and other material particles might also be created and destroyed, given sufficient energy. A pure particle theory could not cope with a situation in which the number of particles is changing. What is more, any notion of individuality is surely exploded if the number of "individuals" can change.

In order to understand this larger realm, we needed to develop a quantum theory of *fields* from the original quantum theory of *particles*. Here Faraday's vision triumphed, though in a way that he would have found passing strange. In 1927, Paul Dirac wanted to unite the quantum theory of particles with Einstein's special theory of relativity. To do so, he found ways of quantizing the electromagnetic field, so that the "particle" of light is a manifestation of that field. To treat both light and electrons on the same footing, he also had to formulate an "electron field" whose manifestations were electrons in different states. These "quantum fields" are very different from the almost palpable lines of force that so impressed Faraday. Mathematically, Dirac's quantum field shows how the number of quanta can be raised or lowered, whether light or electrons. The quantum field represents (for instance) the probability of the creation or annihilation of an electron at a certain space-time point. At every point, the field acts as if it were a dimensionless oscillator whose changing "vibrations" correspond to the creation or destruction of particles.

The unification of quantum theory with relativity led Dirac to propose in 1928 that there was, corresponding to the electron, a corresponding particle of equal but opposite charge and identical mass. Thus the electron has as partner a positron. When both meet, they can annihilate into pure energy, most simply in the form of light. This bold prediction of antimatter was experimentally confirmed in 1932 (see figure 8.5). Stranger still, pairs of particles and antiparticles are continually created and destroyed even in the vacuum, which is filled with such "virtual" pairs. These do not constitute an ether because they are neither substantial nor directly observable; the pairs annihilate so quickly that they escape detection. Even so, they do lead to subtle indirect effects on the motion of observable particles, which tremble measurably in response to the charge fluctuations in the vacuum. This trembling of the orbital electron causes a tiny but measurable shift in the spectrum of hydrogen.

The quantum theory of fields construes electrons and other quanta as manifestations of a deeper level of reality, of quantum fields. Identicality is deeply embedded in quantum field theory, perhaps even more so than in the earlier quantum theory of single particles. Here the best analogy may be a vibrating string, described by classical field theory, giving rise to discrete overtones, each representing a simple whole number division of the string. Think of a violin string vibrating as a whole, giving forth its fundamental tone, say G. Touch the string exactly at its midpoint; one hears a harmonic tone g exactly an octave higher, as the string is divided in the ratio of 1:2. Similarly, one finds another harmonic when the string is touched at the point dividing it in 3:4 ratio, the note c. Between those two points, there is no other similarly resonant overtone. In this way, a continuous string gives rise to discrete, distinct overtones, each always identical. As the bow sweeps over the string, it excites these overtones, giving rise to the violin's characteristic tone color. Analogously, the quantum electromagnetic field gives rise to "overtones" that represent the creation of one, two, three, . . . electrons, all of identical mass and charge. Their identicality stems from the underlying unity of the quantum field, as a vibrating string keeps its overtones in exact proportion to its length. In this way, quantum fields seem to have a deep connection with identicality.

Beyond observable indistinguishability, electrons remain identical in charge even when they are not observable to us, as when they are virtual particles or are so distant that they are not yet visible to us. The maintenance of self-identity, of unaltered fundamental properties, is crucial. As Feynman phrased it, one can think of an antielectron (or positron) as an electron moving backwards in time. Furthermore, the world-line of an electron remains continuous even when (interacting with other quanta, say) it veers backwards in time, appearing in our detectors as a positron. That can be true only if the magnitudes of its charge and mass remain always identical, whatever strange turns its career takes. To make this so, these magnitudes must be deeply embedded in the quantum field. Mathematically, this requirement deeply conditions quantum field theories, which always rest on noncommutation. It underlies some of the most sweeping generalities of that theory, the connection between spin and statistics (namely, that bosons must have integral spin, in units of $h/2\pi$, while fermions have half-integral spin) and the celebrated TCP theorem (which asserts the universal pairing of every particle with an antiparticle).

As strange as Maxwell's field is, the quantum field is stranger still. Perhaps the word "field" here is misleading, because it makes one think of something like Maxwell's field: a number and a direction at every point of space and time, describing the force a unit charge would experience, if it were put there. Mathematically, the quantum field, like the classical field, is defined at every point in space-time, but it has a far different significance. It tells not of certainty but of probability, the probability of the observation of the production of 1, 2, 3, . . . quanta as excitations of the field, not of a single particle. Thus the concepts "particle" and "field" are very misleading. An electron is no more an isolated being than is a particular overtone of the violin string. However, the quantum "field" never manifests itself in any continuous way, unlike the classical field with its smooth waves. The only truly "fieldlike" thing about a quantum field is its strict observance of causality. An effect always happens after its cause, never before, even on the smallest scale observed to date. Influence never jumps but always seems to connect cause and effect continuously, which the smooth "field" functions of the theory ensure. More deeply, this is a consequence of the special theory of relativity, in which the speed of light limits the speed at which causal influence propagates in time.

For all its success, quantum field theory is shadowed by paradox: there are infinities in its predictions, since the theory assumes that space-time is infinitely continuous in order to allow causal influences smooth passage. Despite these infinities, ingenious mathematical techniques can derive perfectly finite predictions from quantum field theory, predictions confirmed by experiment to extremely high precision. Perhaps some other theory will eventually supersede it, though there is no experimental evidence indicating a failure of field theory so far, at the end of the twentieth century. Some hold out great hope for string theory, which posits vibrating primordial "strings," rather than the vibrating dimensionless points assumed in field theory. Yet these strings are themselves presumed to be identical; only their different states of vibration manifest the various observed particles, again recalling the vibrating violin string. Whatever may happen, the notion of identicality is a touchstone that may guide the search for deeper understanding.

These physical ideas reach toward larger philosophical questions of individuality and identity. Throughout this book, I have tried to indicate the continuing dialogue involving these questions and the developments in atomic theory. This dialogue continues today as thoughtful people struggle with these questions in light of modern physics. For instance, consider this updated version of some stories we considered earlier. Imagine that, through biotechnological wizardry, you could be cloned, so that a year from now a new body would be generated from some of your present tissue. Though this goes beyond present ability, this imaginary scenario is uncomfortably close to being possible. Next year, there will be two human beings that look just like you. Say that one of them has most of your present body, while the other has most of your memories. Which of these, then, is *you*? Both? Neither? Only one (but which)? During the intervening year, say that one of them has acted ethically and the other badly. Are you then responsible for the actions of both? If you have a soul, was it also cloned?

This hypothetical story combines the ship of Theseus (the continued identity of one body as it is replaced or even duplicated) with the two Martin Guerres (deciding between two indistinguishable versions of the same person), given a new life through genetic engineering. These vexing problems show the depth of the ancient questions and how alive they remain. Contemporary philosophers are still divided as they discuss how to resolve these difficulties. Increasingly, they find it hard to hold on to the notion that one of the two yous simply has "primitive thisness," the indelible mark that goes beyond anything observable to set *you* off from any other, however closely resembling or even cloned. Think of these two as if they were two Martin Guerres. How would the court now decide between them? Worse still, what would Bertrande do?

If "primitive thisness" is not available as a touchstone, we need other ways to resolve such problems. Philosophers have tended to refer these various difficulties to other human agencies. They say: if I want to know which of the two yous is legally responsible, I should consult a lawyer, not a philosopher. Likewise, a theologian should decide about souls or a physician about bodies. This position seems to stave off some problems by divesting philosophy of its old questions and referring them to other human agencies, presumably better able to deal with them, if only through established practice and custom.

Here we return to identicality. If indeed we can never ultimately know the individuality of the material beings that make up our bodies, we cannot refer our human questions about identity to a physicist, the way we refer our legal problems to a lawyer. After you have been cloned, the physicist will still tell you that there is no physical individuality in the electrons of either of the two yous. Worse still, there never was any such individuality even for outwardly quite different people. This, then, would explain why questions of identity have been so vexed, for microscopic physics is totally incommensurable with the macroscopic objects we live with.

We have only begun to think about the implications of identicality. This realization is truly a philosophical matter, for physics knows nothing of "you" or "I." At least to this extent, the humility of philosophy (or its desperation) does not allow it to forward its problems to other experts. To be sure, professional philosophers do a real service as they carefully discern the subtleties and pitfalls of common opinion, helping us to think more clearly. Yet the true dimensions of philosophy go beyond their expertise, for these issues touch human concerns that we cannot depute to others. A wise judge will consult experts but must finally weigh their testimony critically. If that is so, these problems now devolve on all thinking persons. This is only right and has been long recognized in many fields. The jurist may decide questions of legal right, but cannot answer what is ultimately just. The physician may restore me to life, but cannot answer my question: what should I do with my life? The theologian can show me the implications of my beliefs, but it is I who must believe.

Long ago, Socrates disclaimed all wisdom, except that of knowing his own ignorance. The great philosopher knew that he was not worthy to be called wise. Yet he aspired to love wisdom all the more, knowing how much he lacked it and how deeply he needed it. This love was expressed in endless conversation with everyone he came across, as if such love of wisdom might be as close as we could get to wisdom itself. Let us talk together with the courage, humor, and ardor of Socrates.

In that long conversation, we may find ourselves considering something Plato's follower Plotinus said long ago about "a principle which transcends being," in whose domain one can "assert identity without the affirmation of being." There, "everything has taken its stand for ever, an identity well pleased, we might say, to be as it is. . . . [I]ts entire content is simultaneously present in that identity: this is pure being in eternal actuality; nowhere is there any future, for every then is a now; nor is there any past, for nothing there has ever ceased to be." Individuality and existence in space and time may be masks that our sensibilities impose on the far different face of quantum reality.

Epilogue

Like all genuine questions, the question about identity will never die. Such questions do not have answers, in the sense of a single definitive statement that eliminates the need to ask the question again. Yet that does not mean that talking about such questions is an endless and meaningless game, merely going back and forth over the same positions, more cleverly expressed. Instead, at crucial moments in this long conversation, something emerges that reveals a new truth, perhaps implicit in what has gone before but only now expressed. Because of that insight, everything appears in a new light. Such questions and conversations are living things; they are fascinating because, at any moment, something so compelling may emerge that nothing will be the same again.

At such moments, we realize the narrowness of our preconceptions. Hamlet was right: our philosophy *is* ignorant of most things in heaven and on earth. As Socrates found out, many people do not easily bear the sting of knowing their own ignorance. He tried to show that this was not hurtful, but helpful, purging our false assumptions and narrow opinions. If there is more than we know, eventually we may know more. In this, modern science should continue Socrates' controversial project of examining, testing, and improving human opinions through searching conversation and reasoned inquiry. As a young man, Socrates turned away from natural philosophy because it did not inquire into *why* the world is as it is, and not otherwise. This question haunted Kepler and Einstein; it should continue to haunt us also.

In the story of individuality, contrasting visions confront each other. The individuality of each person and macroscopic object is unique, like Hector's shining helmet. Yet that helmet, like each of us, is made of electrons that blend and merge. The world as we experience it calls us to reconcile these views; vast numbers of identical beings can form structures whose complex configurations give them the appearance of uniqueness. Here we return to the question raised throughout this book, whether the individuality of persons really touches the individuality of things. I may be like the ship of Theseus, a phantom haunting itself, for on the atomic level I have no individuality.

Because of this lack of individuality at the quantum level, it may be important to maintain the sense of strangeness that separates the quantum realm from the human. The mind all too easily assimilates the world into itself, while ignoring what may be truly foreign. If that is so, it is strangely beautiful that human individuality rests on anonymous quanta.

Nevertheless, it is possible to find certain correlates between human experience and this radical merging of individuality. Human individuality is more like a field than isolated, atomic selfhood. If Jean Piaget is correct, each of us begins with no firm limit between our self and the "outside" world. Perhaps one might be able to touch again some of that original wholeness. Lucien Lévy-Bruhl used the phrase *participation mystique* to describe the ardent investment of the self in another object or person, as when a Zuñi dancer becomes the god whose mask he wears: "To exist is to participate. . . . Without participation, one has no existence." Human identity emerges in the intense identification that Lévy-Bruhl calls participation.

In this view, what we feel is not so different from the behavior of our electrons, whose resonance and interference reflects their shared identicality and perhaps ours also. Crowds of people experience such moments of constructive or destructive collective feeling, merged in ecstatic communion or blood lust. But even when solitary, one can experience the truth of Walt Whitman's observation, "I contain multitudes." Sometimes the boundaries of self seem to expand past the limits of self-interest. My sphere of concern grows and shrinks in a way that suggests that I do not end at the edge of my skin, but may extend outward into other persons, other things. At one extreme is the utter destruction of self that is psychosis. Yet even in more tempered moments, the self merges into someone or something beyond itself.

Looking out his window, John Keats felt identified with the sparrow pecking at the ground outside: "I take part in its existence and pick about the Gravel." Poetry and music call us beyond the narrow limits of the self and so does science. Here Spinoza's vision may be the closest: each individual is really a mode of a single universal substance, the field. Feeling the complex interweaving of identity that joins us, each of us may be an aspect of the other. Such attenuation or extension of identity is crucial to love, but it also marks the smallest movement of attention or imagination that stirs the self. If identity is found when it is lost, it may be regained when surrendered. In that case, the strange identity of electrons may give us an image of our own inwardness. Imagine, then, two snowflakes on a winter evening. Consider you and me. Consider us.

Notes

Prologue

Snowflakes: For Johannes Kepler's treatment, see *The Six-Cornered Snowflake*, tr. C. Hardie (Oxford: Oxford University Press, 1966), discussed in my book *Labyrinth: A Search for the Hidden Meaning of Science* (Cambridge: MIT Press, 2000), pp. 87–90. For images of snowflakes, see W. A. Bentley, *Snowflakes in Photographs* (Mineola, NY: Dover, 2000).

Philosophical treatments of identity: There is a vast literature on this subject, of which I will treat only a few examples in this book. For a full overview and many references, see Jorge J. E. Gracia, *Individuality: An Essay on the Foundations of Metaphysics* (Albany: State University of New York Press, 1988). I am deeply indebted to Gracia's treatment, which is well versed in the sources and (what is even rarer) extremely readable. For his clarification of the terms "individuality" (especially understood as "non-instantiability"), "identity," "primitive thisness," and "indistinguishability" (or "indiscernibility"), see pp. 43–46, 38–41, 21–24; for the extension of individuality, see pp. 57–115. A particularly important modern treatment is P. F. Strawson, *Individuals: An Essay in Descriptive Metaphysics* (London: Methuen, 1959). A number of helpful essays are collected in *Identity and Individuation,* ed. Milton K. Munitz (New York: New York University Press, 1971). I have also been helped by E. J. Lowe, *Kinds of Being: A Study of Individuation, Identity and the Logic of Sortal Terms* (Oxford: Blackwell, 1989), which presents a lucid argument for treating individuals always as individuals of a certain kind or "sortal group." This rather Aristotelian framework proves useful in understanding the individuality of quanta.

Chapter 1: Commodity and Sacrament

All translations from Homer are taken from *The Iliad* and *The Odyssey*, tr. Robert Fagles (New York: Viking Penguin, 1990, 1996); the line numbers refer to the book and standard lines of the Greek text.

Hector and the helmet: *Iliad* 6.448–497. Hector is called *phaidimos,* shining or glorious, while the helmet is called *pamphanaousan,* bright-shining or beaming, as of the Sun or burnished metal (6.472–473). Indeed, the Greek word for "appear" or "seem" (*phaino*) is cognate with shining light (*phaos*).

Diomedes and Glaucus: *Iliad* 6.120–150, 234–236.

Currency and early accounting: See Michael Chatfield, *A History of Accounting Thought* (Huntington, NY: Krieger Publishing, 1977) and also A. C. Littleton and B. S. Yamey, *Studies in the History of Accounting* (Homewood, IL: R. D. Irwin, 1956).

Cyclops: *Odyssey* 9.105–566.

Calypso: *Odyssey* 5.13–281. Her name means *the hidden one* or *the one who hides* and Odysseus spends his longest sojourn with her.

Argos: *Odyssey* 17.290–327.

Odysseus and Penelope: *Odyssey* 23.206.

Primitive thisness: a term derived from the Latin *haeccitas,* used by Duns Scotus. See Gracia, *Individuality,* pp. 86, 120 and Robert M. Adams, "Primitive Thisness and Primitive Identity," *Journal of Philosophy* 76:1, 5–26 (1979).

Chapter 2: The Ship of Theseus

Theseus and his ship: *Plutarch's Lives,* tr. John Dryden (New York: Modern Library, 2001), vol. 1, pp. 13–14.

Leucippus and Democritus: *The Presocratics,* ed. Philip Wheelwright (New York: Odyssey Press, 1966), pp. 175–199.

Plato on atomism: *Meno* 76c–76e.

Real: The Latin root of this word is *res,* a thing, indicating concrete substantiality. However, the older source of this concept is the Greek word *alēthēs,* whose root meaning is "unconcealed," literally *a-lēthēs,* that which does not taste *lēthē,* the waters of oblivion. This primal concept denotes the real in the sense of the true, whose opposite is *pseudēs,* the "pseudo" or false. See also Martin Heidegger, *Early Greek Thinking,* tr. D. F. Krell and F. A. Capuzzi (New York: Harper & Row, 1975), pp. 102–123.

Plato and Aristotle on individuality: See Gracia, *Individuality,* pp. 60ff.

Aristotle's joke: "To prove what is obvious by what is not is the mark of a man who is unable to distinguish what is self-evident from what is not." *Physics* 193a6, *The Complete Works of Aristotle,* ed. Jonathan Barnes (Princeton: Princeton University Press, 1984), vol. 1, p. 329.

Socrates as a new Theseus: See Jacob Klein, "Plato's *Phaedo*" in his *Lectures and Essays* (Annapolis, MD: St. John's College Press, 1985), pp. 375–393, and also Ronna Burger, *The* Phaedo: *A Platonic Labyrinth* (New Haven: Yale University Press, 1984), pp. 17, 20, 161.

Modern comments on the ship of Theseus: See David Wiggins, *Identity and Spatio-temporal Continuity* (Oxford: Blackwell, 1967) and *Sameness and Substance* (Oxford: Blackwell, 1980); Hugh S. Chandler, "Rigid Designation," *Journal of Philosophy* 76, 363–369 (1979); Roderick M. Chisholm, "Problems of Identity" in Munitz, *Identity and Individuation,* pp. 3–30.

Chapter 3: Atoms and Monads

Porphyry and Boethius on individuation: Jorge J. E. Gracia, *Introduction to the Problem of Individuation in the Early Middle Ages* (Washington, D.C.: Catholic University of America Press, 1984), pp. 65–121.

Roman Stoics: The citations are given in Max Jammer, *The Conceptual Development of Quantum Mechanics* (New York: McGraw-Hill, 1966), pp. 339–340; for general background, see Andrew G. Van Melsen, *From Atomos to Atom* (New York: Harper, 1960).

Seneca: Thomas G. Rosenmeyer, "Seneca and Nature," *Arethusa* 33, 99–119 (2000).

Hegel on Stoicism: G. W. F. Hegel, *Phenomenology of Spirit,* tr. A. V. Miller (Oxford: Oxford University Press, 1977), p. 121.

Kalām: See Jammer, *Conceptual Development,* p. 341 and his *Concepts of Space* (New York: Harper, 1960), pp. 60–67; Moses Maimonides recounts their theories in *The Guide of the Perplexed,* tr. Shlomo Pines (Chicago: University of Chicago Press, 1963), pp. 194–214, a work that Leibniz read carefully. For a comprehensive treatment, see Alnoor Dhanani, *The Physical Theory of Kalām* (Leiden: E. J. Brill, 1994).

Ancient atomic theory and its early modern revival: See Christoph Lüthy, "The Fourfold Democritus on the Stage of Early Modern Science," *Isis* 91, 443–479 (2000).

Atomism, the Church Fathers, and John Wyclif: See Kurt Lasswitz, *Geschichte der Atomistik vom Mittelalter bis Newton* (Hildesheim: Georg Olms Verlag, 1984), vol. 1, pp. 11–30. Pietro Redondi argued in *Galileo Heretic,* tr. Raymond Rosenthal (London: Penguin, 1983), that Epicureanism and the Eucharist were the hidden central issues in the Galileo case; however, the documents of that case do not explicitly confirm this connection.

Bruno's atomism: See Lüthy, "The Fourfold Democritus," pp. 451–454.

Bacon on atoms: See Graham Rees, "Atomism and 'Subtlety' in Bacon's Natural Philosophy," *Annals of Science* 37, 549–571 (1980) and "Bacon's speculative philosophy" in *The Cambridge Companion to Bacon,* ed. Markku Peltonen (Cambridge: Cambridge University Press, 1996), pp. 121–145 at 132–133, and Lüthy, "The Fourfold Democritus," p. 465.

Gassendi: See L. S. Joy, *Gassendi the Atomist* (Cambridge: Cambridge University Press, 1987), B. Brundell, *Pierre Gassendi* (Dordrecht: Reidel, 1987), Margaret J. Osler, *Divine Will and the Mechanical Philosophy: Gassendi and Descartes on Contingency and Necessity in the Created World* (Cambridge: Cambridge University Press, 1994), and *Gassendi et l'Europe (1592–1792),* ed. Sylvia Murr (Paris: J. Vrin, 1997).

Descartes: *Principles of Philosophy,* tr. V. R. Miller and R. P. Miller (Dordrecht: Kluwer Academic, 1991), pp. 48–49, 284–285 (against atomism); 106–111, 132–136, 141–144 (the cosmic "mill-wheel"). See also E. J. Aiton, *The Vortex Theory of Planetary Motion* (New York: American Elsevier, 1972), pp. 43–55.

Atomism in Boyle and Newton: Robert Hugh Kargon, *Atomism in England from Harriot to Newton* (Oxford: Clarendon Press, 1966).

Newton on atoms: *Opticks* (New York: Dover, 1979), p. 400 and his Rule 3 in *The* Principia: *Mathematical Principles of Natural Philosophy,* tr. I. Bernard Cohen and Anne Whitman (Berkeley: University of California Press, 1999), pp. 795–796.

Individuality in seventeenth-century philosophy: I have been greatly helped by the articles by Udo Thiel on "Individuation" and "Personal Identity" in *The Cambridge History of Seventeenth-Century Philosophy,* ed. Daniel Garber and Michael Ayers (Cambridge: Cambridge University Press, 1998), vol. 1, pp. 212–262, 868–912.

Boyle: *Selected Philosophical Papers of Robert Boyle,* ed. M. A. Stewart (Manchester: Manchester University Press, 1979), pp. 30 (bodies made up of corpuscles), 193 (no such easy way to determine identity of body).

Hobbes on the ship of Theseus: *Elements of Philosophy* (1655) in *The English Works of Thomas Hobbes* (Aalen: Scientia Verlag, 1966), vol. 1, pp. 136–137; see also Chandler, "Rigid Designation."

Locke: marginal note in John Sergeant, *Solid Philosophy Asserted, against the Fancies of the Ideists . . .* (London, 1697; reprint New York: Garland, 1984), p. 258 (what complexion can distinguish two atoms); *An Essay Concerning Human Understanding,* ed. Peter H. Nidditch (Oxford: Oxford University Press, 1975), pp. 330 (continuity of existence of atoms), 338 (two thinking substances may make one person), 110–111 (Castor and Pollux), 107–114 (same man waking or sleeping may be different persons).

Resurrection of the body: Thiel, "Personal Identity," pp. 885–887, 886 (Alexander Ross: God can reunite atoms), 897–899 (Stillingfleet); Boyle: *Selected Papers,* p. 199 (no determinate size to make body the same); Sir Kenelm Digby, *Observations upon Religio Medici . . .* (London, 1644; reprint Menston: Scolar Press, 1973), pp. 85–87.

G. W. Leibniz: *Philosophical Writings,* tr. Roger Ariew and Daniel Garber (Indianapolis: Hackett Publishing Company, 1989), pp. 327–328 (episode in the garden); other passages relating to this matter are on pp. 44, 46–47, 164, 333; *New Essays on Human Understanding,* tr. Peter Remnant and Jonathan Bennett (Cambridge: Cambridge University Press, 1996), pp. 230–231 (atoms are chimerical). See also Bertrand Russell, *A Critical Exposition of the Philosophy of Leibniz* (London: George Allen & Unwin, 1964), pp. 54–66, and Edwin Curley, "Leibniz on Locke on Personal Identity," in *Leibniz: Critical and Interpretive Essays,* ed. Michael Hooker (Minneapolis: University of Minnesota Press, 1982), pp. 302–326.

Hume on identity: *A Treatise of Human Nature,* ed. P. H. Nidditch (Oxford: Oxford University Press, 1978), pp. 204–205 (succession and identity), 207 (the mind), 263–270 (shipwreck and backgammon), 270 ("If I must be a fool").

Chapter 4: Secret Sharers

Martin Guerre: The essential modern retelling is Janet Lewis, *The Wife of Martin Guerre* (Chicago: Swallow Press, 1981). The modern scholar I refer to is Natalie Zemon Davis, *The Return of Martin Guerre* (Cambridge: Harvard University Press, 1983), who considers Bertrande as a self-fashioning woman. There is an interesting critique and response in Robert Finlay, "The Refashioning of Martin Guerre" and Natalie Zemon Davis, " 'On the Lame,' " *American Historical Review* 93:3,

553–571, 572–603 (1988). Jean de Coras's account has been translated with comments by Jeannette K. Ringold and Janet Lewis, "A memorable decision of the High Court of Toulouse," *The Triquarterly* 55, 86–110 (1982).

Montaigne on the Guerre case: See his essay "Of cripples" (1585–1588) in *The Complete Essays of Montaigne,* tr. Donald M. Frame (Stanford: Stanford University Press, 1958), pp. 784–792.

Comedy of Errors: I.ii.35–38. This imagery of water and melting recurs in other passages concerning dissolution of identity or reality; see H. F. Brooks, "Themes and Structure in *The Comedy of Errors,*" in *Early Shakespeare,* ed. J. R. Brown and Bernard Harris (Stratford-upon-Avon: Stratford-upon-Avon Studies 3, 1961), pp. 55–71.

Doubles: Otto Rank, *The Double* (New York: New American Library, 1979) and Robert Rogers, *A Psychoanalytic Study of the Double in Literature* (Detroit: Wayne State University Press, 1970), which discusses many examples.

Fyodor Dostoyevsky: *The Double,* tr. Constance Garnett (Mineola, NY: Dover, 1997), pp. 43 (he began to doubt his own existence), 67 (treated like a rag), 87 (a terrible multitude of duplicates).

Joseph Conrad: *Heart of Darkness and The Secret Sharer* (New York: New American Library, 1950), pp. 27–28, 30–31 (he appealed to me), 50 (near insanity), 52 (you understand).

Chapter 5: Distinguishability and Paradox

The historical background of Leibniz's Principle: For the medieval background, see Gracia, *Individuation;* for the later developments see his *Suárez on Individuation* (Milwaukee: Marquette University Press, 1982), and his *Individuals,* pp. 196–200. Regarding Leibniz, see also Leonard J. Eslick, "Aristotle and the Identity of Indiscernibles," *Modern Schoolman* 36, 279–287 (1959); Thomas P. McTighe, "Nicholas of Cusa and Leibniz's Principle of Indiscernibility," *Modern Schoolman* 42, 33–46 (1964); Lawrence B. McCullough, *Leibniz on Individuals and Individuation* (Dordrecht: Kluwer, 1996); Roberto Torretti, *The Philosophy of Physics* (Cambridge: Cambridge University Press, 1999), pp. 98–101; and my essay, "Leibniz and the Leaves: Beyond Identity," *Philosophy Now* 18–21 (December 2000/January 2001).

The Roman concept of personhood: Edward Gibbon, *The History of the Decline and Fall of the Roman Empire* (New York: Penguin,

1994), vol. 2, pp. 806–811 (the *patria potestas*), 789–790 (manumission), discussed by Philip LeCuyer, "Persona" (unpubl. lecture).

Persona and personhood: See the helpful essay by Bruce Venable, "The Name and Nature of the Person," in *Essays in Honor of Robert Bart,* ed. Cary Stickney (Santa Fe, NM: St. John's College, 1993), pp. 260–274.

Aquinas on angels: See his *Summa Theologica,* First Part, Question 50, Article 4.

Davy, Herschel, and atomism: See *The Atomic Debates,* ed. W. H. Brock (Leicester: Leicester University Press, 1967), pp. 1–11.

Dalton on the identity of atoms: See John Dalton, *A New System of Chemical Philosophy* (New York: Citadel Press, 1964), p. 113. For other reflections on Dalton's argument, see Abraham Pais, *Inward Bound* (Oxford: Oxford University Press, 1986), pp. 72 (Maxwell), 122 (Thomson), 146 (Soddy); for Maxwell's extension of this argument see the present volume, chapter 7. For multiple proportions, see John Bradley, *Before and After Canizzaro* (North Ferriby: J. Bradley, 1992), pp. 32–39.

Boltzmann on distinguishability: See Ludwig Boltzmann, *Theoretical Physics and Philosophical Problems* (Dordrecht: D. Reidel, 1974), pp. 228–231 and Hans Reichenbach, *Philosophic Foundations of Quantum Mechanics* (Mineola, NY: Dover, 1998), p. 1.

The horse and stone "paradox": Newton's *Principia,* p. 417; for a nice account of grappling with the paradox, see Jed Z. Buchwald, *From Maxwell to Microphysics: Aspects of Electromagnetic Theory in the Last Quarter of the Nineteenth Century* (Chicago: University of Chicago Press, 1985), p. 3: "I now saw, far from *nothing* moving, *everything* moves since the equal and opposite forces act on different bodies."

Entropy: See James Clerk Maxwell, *Theory of Heat,* ed. Peter Pesic (Mineola, NY: Dover, 2001), pp. 139–162; for the cultural significance of Carnot and entropy, see George Steiner, *After Babel* (New York: Oxford University Press, 1975), pp. 152–156.

Maxwell's kinetic theory of gases: See Maxwell, *Theory of Heat,* pp. 298–324.

Gibbs's Paradox: For detailed treatment and references, see my paper, "The principle of identicality and the foundations of quantum theory. I. The Gibbs paradox," *American Journal of Physics* 59, 971–974 (1991). Ehrenfest and Trkal (1920) first pointed out that entropy does not have a clear definition unless changes of the number of particles are included; this allows resolution of the paradox. See H. Grad,

"The many faces of entropy," *Communications in Pure and Applied Mathematics* 14, 323–353 (1961) and N. G. van Kampen, "The Gibbs paradox," in *Essays in Theoretical Physics, in Honour of Dirk ter Haar*, ed. W. E. Parry (Oxford: Pergamon, 1984), pp. 303–312. For Gibbs's own account, see his *Elementary Principles in Statistical Mechanics* (New York: Dover, 1960), pp. 187–207.

Permutation of indistinguishable atoms: Consider first a deck of 52 cards from which we will draw cards, one at a time. For the first card, there are 52 possibilities; for each of these, there are 51 cards left to be drawn as a second card, so that there are 52×51 possible draws of two cards. Similarly, there are $52 \times 51 \times 50$ draws of three successive cards, and $52 \times 51 \times 50 \times \cdots \times 3 \times 2 \times 1$ draws of 52 cards (called "52!" or "52 factorial"). If there are N identical atoms, an exactly parallel argument shows that there are $N! = 1 \times 2 \times 3 \times 4 \times \cdots \times N$ possible permutations that are indistinguishable, which Gibbs divides out.

Chapter 6: The Fields of Light

For further details and references, see my essay "The Fields of Light," *St. John's Review* 38:3, 1–16 (1988–1989), which this chapter corrects in certain points.

The origins and development of field theory: Albert Einstein and Leopold Infeld, *The Evolution of Physics* (New York: Simon and Schuster, 1961), pp. 125–153; Max Jammer, *Concepts of Force: A Study in the Foundations of Dynamics* (Cambridge: Harvard University Press, 1957); Mary B. Hesse, *Forces and Fields* (Totowa, NJ: Littlefield, Adams, & Co., 1965); L. Pearce Williams, *The Origins of Field Theory* (Langham: University Press of America, 1980); P. M. Harman, *Energy, Force, and Matter: The Conceptual Development of Nineteenth-Century Physics* (Cambridge: Cambridge University Press, 1982).

Inverse-square forces: In three-dimensional space, the surface area of a sphere increases as the square of its radius. If one imagines the lines of force from an inverse-square law, their density decreases inversely as the square of the distance from the source. But since the area of the sphere increases exactly in proportion as the density of field lines decreases, the total number of field lines will remain the same, no matter what the radius of the sphere. This corresponds to the source having the same apparent strength ("mass" or "charge," when the force is gravitational or electric, respectively), no matter from where it is observed. Known as Gauss's Law, this expresses the conservation of mass (or charge). Note also that if space were not three-dimensional, this argument would fail, implying an intimate connection between three-dimensionality of space and inverse-square law forces. See also Hans Reichenbach, *The Philosophy of Space and Time* (New York: Dover, 1958), pp. 273–283.

Newton on mediation and action at a distance: See Richard S. Westfall, *Force in Newton's Physics* (New York: American Elsevier, 1971) and *Never at Rest: A Biography of Isaac Newton* (Cambridge: Cambridge University Press, 1980), pp. 505–506 (the letter to Bentley).

Other quotations from Newton: *Opticks,* pp. 362–363 (wave theory of light), 368 (against Huygens's ether), 352 (Newton's rarefied ether); *Principia,* p. 944 (laws governing electric spirit).

Writings by Faraday: *Experimental Researches in Electricity* (New York: Dover, 1965), cited as ERE; *The Selected Correspondence of Michael Faraday,* ed. L. Pearce Williams (Cambridge: Cambridge University Press, 1971), cited as SC.

Writings about Faraday: John Tyndall, *Faraday as a Discoverer* (New York: Thomas Y. Crowell, 1961); L. Pearce Williams, *Michael Faraday* (New York: Simon & Schuster, 1971); Geoffrey Cantor, *Michael Faraday, Sandemanian and Scientist: A Study in Science and Religion in the Nineteenth Century* (New York: St. Martin's Press, 1991); Howard J. Fisher, *Faraday's* Experimental Researches in Electricity: *Guide to a First Reading* (Santa Fe, NM: Green Lion Press, 2001).

Quotations from Faraday: SC 2.882 (Maxwell's letter to Faraday), SC 1.134 (Faraday's letter to Ampère), SC 2.864 (Faraday's letter to Maxwell).

Faraday and Boscovich: See Jammer, *Concepts of Force,* pp. 170–179, 181–185, who argues that Faraday endorsed Boscovich's ideas, at least temporarily, but notes that by 1855 Faraday wrote more cautiously that "I give the lines of force only as representations of the magnetic power, and do not profess to say to what physical ideas they may hereafter point, or into what they will resolve themselves" (p. 184). See also Tyndall, *Faraday as a Discoverer,* p. 148; Mary B. Hesse, "Action at a distance in classical physics," *Isis* 46, 337 (1965); and Williams, *Faraday,* pp. 73–89, who represents Faraday as committed to Boscovich's point-atoms by the 1820s.

Faraday on lines of force: "Thoughts on Ray Vibrations" in ERE 3.451 (radiation as vibration of lines of force); "On the Physical Lines of Magnetic Force" in ERE 3.438–443 (physical reality of lines of force); ERE 3.435–436 (magnets as habitations of lines of force).

"I see a motion, not a form": Janet Lewis, "Geometries" in her *Poems Old and New 1918–1978* (Athens, OH: Swallow Press, 1981), p. 107.

Writings by Maxwell: *The Scientific Papers of James Clerk Maxwell,* ed. W. D. Niven (New York: Dover, 1965), cited as SP; *A Treatise on Electricity and Magnetism* (New York: Dover, 1954), cited as TEM; *The Scientific Letters and Papers of James Clerk Maxwell,* ed. P. M. Harman (Cambridge: Cambridge University Press, 1990, 1995), cited as LP.

Writings about Maxwell: C. W. F. Everitt, *James Clerk Maxwell: Physicist and Natural Philosopher* (New York: Scribner's Sons, 1975); Martin Goldman, *The Daemon in the Aether: The Story of James Clerk Maxwell* (Edinburgh: Paul Harris, 1983); John Hendry, *James Clerk Maxwell and the Theory of the Electromagnetic Field* (Bristol: Adam Hilger, 1986); Thomas K. Simpson, "Maxwell's *Treatise* and the Restoration of the Cosmos," in *The Great Ideas Today* (Chicago: Encyclopaedia Brittanica, 1986), pp. 218–267; P. M. Harman, *The Natural Philosophy of James Clerk Maxwell* (Cambridge: Cambridge University Press, 1998), pp. 71–90 (field theory), 162–187 (ether and molecules), 195–196 (Faraday's conception of matter); Thomas K. Simpson, *Figures of Thought: A Study of Maxwell's* Treatise (Santa Fe, NM: Green Lion Press, forthcoming).

Quotations from Maxwell: "Faraday," in SP 2.359–360; TEM 1.viii–x (Faraday); SP 2.448 (the "vulgar opinion" about impenetrable atoms); Maxwell, *Theory of Heat,* pp. 330–332 ("intermediate links" in biology and physics); SP 2.374–377, 480–484 (equality of molecular vibrations); SP 2.376–377, LP 1.427, 670 (the limitations of physical science).

Spectra: For a brief account, see Abraham Pais, *Niels Bohr's Times, in Physics, Philosophy, and Polity* (Oxford: Clarendon Press, 1991), pp. 139–143; for full details, see W. McGucken, *Nineteenth Century Spectroscopy* (Baltimore: Johns Hopkins University Press, 1969) and J. B. Hearnshaw, *The Analysis of Starlight: One Hundred and Fifty Years of Astronomical Spectroscopy* (Cambridge: Cambridge University Press, 1986).

Chapter 7: Entanglement

For fuller historical details and references, see my paper, "The principle of identicality and the foundations of quantum theory. II. The role of identicality in the formation of quantum theory," *American Journal of Physics* 59, 975–978 (1991).

For general background, see Max Jammer, *The Conceptual Development of Quantum Mechanics* (New York: McGraw-Hill, 1966), pp. 1–28; Armin Hermann, *The Genesis of Quantum Theory (1899–1913)* (Cambridge: MIT Press, 1971); Hans Kangro, *Early History of Planck's Radiation Law* (London: Taylor & Francis, 1976); Jagish Mehra and Helmut Rechenberg, *The Historical Development of Quantum Theory* (New York: Springer-Verlag, 1982), vol. 1, pt. 1, pp. 23–99, and pt. 2, pp. 613–639; Christa Jungnickel and Russell McCormmach, *Intellectual Mastery of Nature* (Chicago: University of Chicago Press, 1986), vol. 2, pp. 248–252, 260–268; Thomas S. Kuhn, *Black-Body Theory and the Quantum Discontinuity 1894–1912* (Chicago: University of Chicago Press, 1987), pp. 3–71.

History of the electron: See R. Millikan, *The Electron* (Chicago: University of Chicago Press, 1963) and Pais, *Inward Bound,* pp. 67–92, who outlines the many complexities behind the final discovery. These include an important technical component; Pais also notes that pressures of about .01 mm of mercury are needed to perform precise experiments on the cathode ray beam (760 mm is normal atmospheric pressure). Regarding Millikan's sole credit for the experiment, see Harvey Fletcher, "My work with Millikan on the oil-drop experiment," *Physics Today* 35:6, 43–47 (1982).

Planck's "search for the absolute": Max Planck, *Scientific Autobiography and Other Papers* (New York: Philosophical Library, 1949), p. 47.

Planck and the introduction of the quantum: I am much indebted to the work of Martin J. Klein, who clarified the development of Planck's ideas; see his essays "Max Planck and the Beginnings of the Quantum Theory," *Archive for the History of Exact Sciences* 1, 459–479 (1960) and "Thermodynamics and Quanta in Planck's Work," *Physics Today* 19:11, 23–32 (1966); for Planck's "act of desperation," see p. 27. See also Olivier Darrigol, "Statistics and combinatorics in early quantum theory," *Historical Studies in Physical Sciences* 19:1, 17–80 (1988).

Kuhn argued in his *Black-Body Theory* that Einstein rather than Planck truly introduced the quantum hypothesis; I hope my account helps clarify what Planck did in the realm of recognizing the identicality of quanta, at least in their novel counting. Kuhn's arguments emphasize Planck's hesitations and ambivalence in introducing quantization. For responses to Kuhn, see Martin J. Klein, Abner Shimony, and Trevor J. Pinch, "Paradigm Lost? A Review Symposium" *Isis* 70, 429–440 (1979) and Peter Galison, "Kuhn and the Quantum Controversy," *British Journal for the Philosophy of Science* 22, 71–85 (1981). According to Eugene Wigner, "that Planck believed in the quantum jump is evident," yet Planck did not believe "in the details in any derivation of [his equation], and this was natural since the physics of that time was full of contradictions"; cited in *Some Strangeness in the Proportion,* ed. Harry Woolf (Reading, MA: Addison-Wesley, 1980), p. 194. Moreover, Einstein and Bohr both considered Planck the discoverer of the quantum; see Pais, *Niels Bohr's Times,* pp. 82–87. Kuhn replied to his critics in the afterword to his *Black-Body Theory,* pp. 349–370.

Boltzmann: See Carlo Cercignani, *Ludwig Boltzmann: The Man Who Trusted Atoms* (Oxford: Oxford University Press, 1998), pp. 120–133 (statistical interpretation of entropy), 202–211 (Boltzmann, Planck, and the energeticists) and David Lindley, *Boltzmann's Atom* (New York: Free Press, 2001).

Einstein, Perrin, Smoluchowski, and atomism: See Abraham Pais, *"Subtle is the Lord . . .": The Science and the Life of Albert Einstein* (Oxford: Oxford University Press, 1982), pp. 79–107, and Jean Perrin, *Atoms* (Woodbridge, CT: Ox Bow Press, 1990). These issues will be reexamined in my book *Sky Blue* (in preparation).

Planck's later (1909) account of the quantum: Max Planck, *Eight Lectures in Theoretical Physics,* ed. Peter Pesic (Mineola, NY: Dover, 1998), pp. 1–20, 41–57, 87–96; for further background, see pp. viii–xiii. Note that, contrary to many accounts by others, Planck did *not* use his imaginary oscillator or radio to model the atoms of the oven, but only to give an idealized probe of the electromagnetic radiation they emit; see pp. 81–96.

Quantum theory: For a helpful general overview, see Abner Shimony, "Conceptual foundations of quantum mechanics," in *The New Physics,* ed. Paul Davies (Cambridge: Cambridge University Press, 1992), pp. 373–395. See also Richard C. Henry, "Quantum mechanics made transparent," *American Journal of Physics* 58, 1087–1100 (1990). Especially concerning philosophical aspects, see Hans Reichenbach, *Philosophic Foundations of Quantum Mechanics* (Mineola, NY: Dover, 1998) and Howard Stein, "On the Conceptual Structure of Quantum Mechanics," in *Paradigms and Paradoxes: The Philosophical Challenge of the Quantum Domain* (Pittsburgh: University of Pittsburgh Press, 1972), pp. 367–438.

Bohr on quantum theory: See Pais, *Niels Bohr's Times*, pp. 295–323.

Dyson on quantum theory: Freeman Dyson, "Innovation in Physics," *Scientific American* 199:3, 74–82 (1958).

Feynman on quantum theory: Richard Feynman, Robert Leighton, and Matthew Sands, *The Feynman Lectures in Physics* (Reading, MA: Addison-Wesley, 1965), vol. 3, p. 1-10.

Loss of individuality and Natanson, Ehrenfest, Einstein, and Bose: See Pais, *"Subtle is the Lord . . . ,"* pp. 423–434, and my "The principle of identicality II."

Weyl on individuality: Hermann Weyl, *The Theory of Groups and Quantum Mechanics* (New York: Dover, 1950), p. 241. Weyl thought that electrons still obeyed Leibniz's Principle, but in reverse: if Mike and Ike are indeed indistinguishable, they are thus identical. (In contrast, Leibniz held that if Mike and Ike are different individuals, they must be distinguishable.)

Schrödinger and de Broglie on individuality: Erwin Schrödinger, *The Interpretation of Quantum Mechanics* (Woodbridge, CT: Ox Bow Press, 1995), p. 32; Louis de Broglie, *The Revolution in Physics* (New York: Greenwood Press, 1969), pp. 280–281.

Chapter 8: Identicality

Material from this chapter appeared in another form in my essay on "Quantum Identity," *American Scientist* 90:3, 262–267 (2002).

Spin: See Sin-itiro Tomonaga, *The Story of Spin*, tr. Takeshi Oka (Chicago: University of Chicago Press, 1997).

Millikan on the exactness of the electron's properties: *The Electron*, p. 70.

Identicality: I have given detailed mathematical arguments that this concept is at the heart of quantum theory in "Identity and the foundations of quantum theory," *Foundations of Physics Letters* 13:1, 55–69 (2000). My argument expresses the principle of identicality in terms of four postulates (equality, reversibility, base equivalence, and indistinguishability) and then makes use of an important theorem by D. Fivel, "How interference effects in mixtures determine the rules of quantum mechanics," *Physical Review A* 50, 2108–2119 (1994). To the best on my knowledge, the term "identicality" was first used in my earlier papers on "The principle of identicality."

Philosophical questions of quantum identity: I can mention only a few of the many important works. W. v. O. Quine, "Whither Physical Objects?" in *Essays in Memory of Imre Lakatos,* ed. R. S. Cohen, P. K. Feyerabend, and M. W. Wartofsky (Dordrecht: Reidel, 1976); B. van Fraassen, "The Problem of Indistinguishable Particles," in *Science and Reality: Recent Work in the Philosophy of Science,* ed. J. T. Cushing, C. F. Delaney, and G. M. Gutting (Notre Dame, IN: University of Notre Dame Press, 1984); D. Shapere, "The Origin and Nature of Metaphysics," *Philosophical Topics* 18, 163–174 (1990); B. van Fraassen, *Quantum Mechanics: An Empiricist View* (Oxford University Press, Oxford, 1991), pp. 432–433, 479–480; M. Redhead and P. Teller, "Particles, Particle Labels, and Quanta," *Foundations of Physics* 21, 43–62 (1991) and "Particle Labels and the Theory of Indistinguishable Particles in Quantum Mechanics," *British Journal for the Philosophy of Science* 43, 201–218 (1992); Jeremy Butterfield, "Interpretation and Identity in Quantum Theory," *Studies in the History and Philosophy of Science* 24, 443–476 (1993); N. Huggett, "Identity: Quantum Mechanics and Common Sense," *The Monist* 80:1, 118–130 (1997); S. French and D. Krause, "Vague Identity and Quantum Non-Individuality," *Analysis* 55, 20–26 (1995); and S. French, "Identity and Individuality in Classical and Quantum Physics," *Australasian Journal of Philosophy* 67, 432–446 (1989). For valuable collections of philosophical essays on identity in quantum theory, see *Philosophical Consequences of Quantum Theory: Reflections on Bell's Theorem,* ed. J. T. Cushing and E. McMullin (Notre Dame, IN: University of Notre Dame Press, 1989) and *Interpreting Bodies,* ed. Elena Castellani (Princeton: Princeton University Press, 1998).

Leibniz's Principle in the light of quantum theory: There has been a long controversy over this, which seems to have led to the consensus that the Principle no longer holds in quantum theory. See H. Margenau, "The Exclusion Principle and its Philosophical Importance," *Philosophy of Science* 11, 187–208 (1944); Hans Reichenbach, *Philosophy of Space and Time,* pp. 269–273 and *The Direction of Time* (Berkeley: University of California Press, 1956), sect. 26; H. Post, "Individuality and Physics," *The Listener* 70, 534–537 (1963) (which introduced the term "transcendental individuality" for "primitive thisness"); A. Cortes, "Leibniz's Principle of the Identity of Indiscernibles: A False Principle," *Philosophy of Science* 45, 466–470 (1978); R. L. Barnette, "Does Quantum Mechanics Disprove the Principle of the Identity of Indiscernibles?" *Philosophy of Science* 45, 491–505 (1978); A. Ginsberg, "Quantum Theory and the Identity of Indiscernibles Revisited," *Philos-*

ophy of Science 48, 487–491 (1981), pp. 153–172; S. French and M. Redhead, "Quantum Physics and the Identity of Indiscernibles," *British Journal for the Philosophy of Science* 39, 233–246 (1988); and S. French, "Why the Principle of the Identity of Indiscernibles Is Not Contingently True Either," *Synthese* 78, 141–166 (1989).

"Non-supervenient relations": See P. Teller, "Relational Holism and Quantum Mechanics," *British Journal for the Philosophy of Science* 37, 71–81 (1986) and "Relativity, Relational Holism, and the Bell Inequalities," in Cushing and McMullin, *Philosophical Consequences of Quantum Theory;* S. French, "Individuality, Supervenience, and Bell's Theorem," *Philosophical Studies* 55, 1–22 (1989); S. French and D. Krause, "Vague Identity"; S. French, "Withering Away of Physical Objects," in *Interpreting Bodies,* pp. 93–113. For an argument that classical physics also does not unambiguously support a single metaphysics of identity, see N. Huggett, "Atomic metaphysics," *Journal of Philosophy* 96, 5–24 (1999).

Newton and Occam's razor: *Principia,* p. 794.

Two-path experiments: See Richard P. Feynman, *The Feynman Lectures on Physics,* vol. 3, pp. 1-10–1-11, or (less technically) Feynman's book *The Character of Physics Law* (Cambridge: MIT Press, 1965), pp. 127–148; my paper "Identity" also discusses this example. My treatment here follows the setup of the experiments of J. Summhammer et al., "Direct observation of fermion spin superposition by neutron interferometry," *Physical Review A* 27, 2523–2532 (1983) and H. Rauch and S. A. Werner, *Neutron Interferometry: Lessons in Experimental Quantum Mechanics* (Oxford: Clarendon Press, 2000), as described by A. J. Leggett, "Quantum Mechanics at the Macroscopic Level," in *The Lesson of Quantum Theory,* ed. Jorrit de Boer, Erik Dal, and Ole Ulfbeck (Amsterdam: North-Holland, 1986), pp. 35–57.

Interference of atoms and molecules: See Markus Arndt et al., "Wave-particle duality of C_{60} molecules," *Nature,* 401, 680–682 (14 October 1999), which envisages extending such interference experiments to larger macromolecules, even small viruses.

Cases when quanta can be distinguished: In general, when the wavefunctions of two quanta do not overlap, it is possible to distinguish them in space and time, as in the example of the two Millikan apparatuses given in the text. However, this does not imply any ability to mark or truly to distinguish the quanta in their future course. See Leonard Schiff, *Quantum Mechanics* (New York: McGraw-Hill, 1968 [third edition]), pp. 367–368, and also R. Mirman, "Experimental Meaning of the Concept of Identical Particles," *Nuovo Cimento B* 18, 110–122 (1973).

Particle tracks: For a historical case study, see Brigitte Falkenburg, "The Analysis of Particle Tracks: A Case for Trust in the Unity of Physics," *Studies in the History and Philosophy of Modern Physics* 27, 337–371 (1996).

Images of the atoms: For an excellent popular account of STM images and single trapped atoms (including many striking images), see Hans Christian von Baeyer, *Taming the Atom: The Emergence of the Visible Microworld* (New York: Random House, 1992).

Born and the two levels of quantum theory: See Pais, *Inward Bound,* pp. 258–267.

Einstein's objections to quantum theory: See Pais, *"Subtle is the Lord . . . ,"* pp. 435–459, and also Don Howard, "Einstein on Locality and Separability," *Studies in the History and Philosophy of Science* 16, 171–201 (1985) and "A Peek behind the Veil of Maya: Einstein, Schopenhauer, and the Historical Background of the Conception of Space as a Ground for the Individuation of Physical Systems," in *The Cosmos of Science: Essays of Exploration,* ed. J. Earman and J. D. Norton (Pittsburgh: University of Pittsburgh Press, 1997), pp. 120ff. I discuss Einstein's objections in another context in my *Labyrinth,* chapter 10. For Bohr's response, see Pais, *Niels Bohr's Times,* pp. 316–320, 425–436.

Bell on quantum theory: See J. S. Bell, *Speakable and Unspeakable in Quantum Mechanics* (Cambridge: Cambridge University Press, 1993).

Probability and amplitudes: Strictly, since the amplitude is a complex number, the probability is its absolute square (the amplitude times its complex conjugate). A simple argument shows the need for complex numbers: if there is interference in the quantum case, the probabilities must at certain points be reduced, compared to classical particles. But such reduction requires a *negative* quantity in the probability and that requires an imaginary quantity in the amplitude, since the amplitude is the square root of the probability. Strictly, one can use real-valued matrices instead of complex numbers, but in either case, the square has to be negative. For the significance and history of complex quantities in quantum theory, see Chen Ning Yang, "Square root of minus one, complex phases and Erwin Schrödinger," in *Schrödinger,* ed. C. W. Kilmister (Cambridge: Cambridge University Press, 1988), pp. 53–64, and W. E. Baylis, J. Huscilt, and Jiansu Wei, "Why *i*?" *American Journal of Physics* 60, 788–797 (1992).

Bose-Einstein statistics: See M. Delbrück, "Was Bose-Einstein Statistics Arrived at by Serendipity?" *Journal of Chemical Education* 57, 467–470 (1980).

Symmetry, bosons, and fermions: See Sam Treiman, *The Odd Quantum* (Princeton: Princeton University Press, 1999), pp. 149–172; Dennis Dieks, "Quantum Statistics, Identical Particles, and Correlations," *Synthese* 82, 127–155 (1990); and Nick Huggett, "On the Significance of Permutation Symmetry," *British Journal for the Philosophy of Science* 50, 325–347 (1999).

Anyons: In certain two-dimensional systems, there are other possibilities besides normal bosons and fermions; see Frank Wilczek, "Anyons," *Scientific American* 264:5, 58–65 (1991).

Pauli exclusion principle: For Wolfgang Pauli's account, see his Nobel lecture, "Exclusion Principle and Quantum Mechanics" in his *Writings on Physics and Philosophy*, tr. Robert Schlapp (Berlin: Springer-Verlag, 1994), pp. 165–181, where he noted: "Already in my original paper I stressed the circumstance that I was unable to give a logical reason for the exclusion principle or to deduce it from more general assumptions. I had always the feeling and I still have it today, that this is a deficiency." This may give additional weight to the assumption of identicality as a fundamental postulate. Regarding the deduction of the principle, see I. G. Kaplan, "The exclusion principle and indistinguishability of identical particles in quantum mechanics," *Soviet Physics Uspekhi* 18, 988–994 (1976) and Yoel Tikochinsky and Dan Shalitin, "Simple proof of the symmetrization postulate in quantum mechanics," *American Journal of Physics* 58, 78–79 (1989). See also Y. Shadmi, "Teaching the exclusion principle with philosophical flavor," *American Journal of Physics* 46, 844–848 (1978).

Building the periodic table: See Treiman, *Odd Quantum*, pp. 166–168.

Chemical implications of identicality: See Roald Hoffmann, *The Same and Not the Same* (New York: Columbia University Press, 1995), pp. 32–55, and Jean Jaques, *The Molecule and its Double*, tr. Lee Scanlon (New York: McGraw-Hill, 1993).

Chapter 9: The Silence of the Sirens

Einstein and the Olympian Academy: See my book *Labyrinth*, chapter 10.

Kant on individual identity: See his *Critique of Pure Reason*, tr. Norman Kemp Smith (New York: St. Martin's Press, 1965), pp. 278–284; for Kant's metaphor of the island of truth, see p. A236 (B295); for the example of two indistinguishable drops of water, see p. A263 (B319).

On this argument, see Ian Hacking, "The Identity of Indiscernibles," *Journal of Philosophy* 72:9, 249–256 (1975), who argues that Kant's counterexample (or indeed any example in an imaginary or possible world) is inconclusive. Steven French adds a helpful reply in "Hacking Away at the Identity of Indiscernibles: Possible Worlds and Einstein's Principle of Equivalence," *Journal of Philosophy* 92, 455–466 (1995).

Newton's melancholy: See my *Labyrinth,* chapter 9.

Einstein and Spinoza: See my essay "Einstein and Spinoza: Determinism and Identicality Reconsidered," *Studia Spinozana* 12, 193–201 (1996). Einstein writes of "Spinoza's precarious God" in his *Letters to Maurice Solovine,* ed. M. Solovine (New York: Philosophical Library, 1987), p. 129 (letter of July 30, 1951).

Bohr's argument with Einstein: See Niels Bohr, "Discussion with Einstein on Epistemological Problems in Atomic Physics" in Paul Arthur Schilpp, *Albert Einstein, Philosopher-Scientist* (New York: Harper, 1959), vol. 1, pp. 201–241.

Einstein and Kafka: For further discussion and references, see my essay "Before the Law: Einstein and Kafka," *Literature and Theology* 8:2, 174–192 (1994). Quotations from Kafka come from Gustav Janouch, *Conversations with Kafka* (New York: New Directions, 1971), pp. 97 (everyone a labyrinth), 63 (God must remain hidden), 63 (distorted image), 189 (achieving greatness), 71 (iron fist of technology), 15 (intellectual labor). For the cultural significance of identity, see George Steiner, *Language and Silence* (New York: Athenaeum, 1967), pp. 386–387. The story about Einstein and *The Trial* is told in *The Kafka Problem,* ed. Angel Flores (New York: Gordian Press, 1975), p. xi, naming Thomas Mann as the friend who lent Einstein the book.

"The Silence of the Sirens": Franz Kafka, *The Complete Stories* (New York: Schocken Books, 1976), pp. 430–432.

Kafka and individuation: See Walter H. Sokel, *Franz Kafka* (New York: Columbia University Press, 1966) and Erich Heller, *Franz Kafka* (New York: Viking Press, 1974), pp. 28–29.

Penal colony: "In the Penal Colony," in Kafka, *Complete Stories,* pp. 140–167.

Schopenhauer and music: See Arthur Schopenhauer, *The World as Will and Representation* (New York: Dover, 1969), vol. 1, pp. 131, 255–267 and Rudiger Safranski, *Schopenhauer and the Wild Years of Philosophy,* tr. Ewald Osers (Cambridge: Harvard University Press, 1990), pp. 268–269, 343.

Gregor Samsa: "The Metamorphosis," in Kafka, *Complete Stories,* pp. 130–131.

Einstein as plumber: See *Einstein on Peace,* ed. Otto Nathan and Heinz Norden (New York: Schocken Books, 1968), p. 613. Kafka's remark about crafts comes from Janouch, *Conversations with Kafka,* p. 15.

Born on Einstein: *The Born-Einstein Letters,* ed. Max Born (New York: Walker and Co., 1971), p. 130.

Einstein on marriage: Abraham Pais, *Einstein Lived Here* (New York: Oxford University Press, 1994), pp. 1–29; 25 (twice failed rather shamefully).

Chapter 10: Beyond Being

Validity of Pauli exclusion principle: See, for instance, D. Kekez, A. Ljubičić, and B. A. Logan, "An upper limit to violations of the Pauli exclusion principle," *Nature* 348, 22 (1990). The experiment tested the exclusion principle by setting a severe lower limit on the probability of a nuclear beta-decay process that would violate it.

Millikan on counting electrons: *The Electron,* p. 71.

"Real numbers": I explore the emergence and consequences of the modern concept of numbers (especially in relation to equations and their solvability) in my book *Abel's Proof: An Essay on the Sources and Meaning of Mathematical Unsolvability* (Cambridge: MIT Press, 2003).

"Non-ordinal numbers": These have been treated mathematically as "quasi-sets" by Décio Krause, "On a Quasi-Set Theory," *Notre Dame Journal of Formal Logic* 33, 402–411 (1992) and D. Krause and S. French, "A Formal Framework for Quantum Non-Individuality," *Synthese* 102, 195–214 (1995). For a different approach to these quasisets, see M. L. Dalla Chiara and G. Toraldo di Francia, "Individuals, Kinds, and Names in Physics," in *Bridging the Gap: Philosophy, Mathematics, and Physics,* ed. G. Corsi, M. L. Dalla Chiara, and G. C. Ghirardi (Dordrecht: Kluwer Academic, 1993), pp. 261–283, and Steven French and Décio Krause, "The Logic of Quanta," in *Conceptual Foundations of Quantum Field Theory,* ed. Tian Yu Cao (Cambridge: Cambridge University Press, 1999), pp. 324–342; a comparison of the approaches is given by R. Giuntini, "Quasiset Theories for Microobjects: A Comparison," in *Individual Bodies.*

Schrödinger and configuration space: See Jon Dorling, "Schrödinger's original interpretation of the Schrödinger equation: A rescue attempt," in *Schrödinger,* pp. 16–40, and also Tomonaga, *Story of*

Spin, pp. 95–112, 221–222. Tomonaga emphasizes that, while the ordinary wave function in general cannot exist in three-dimensional space, the "second-quantized" wave function *can*. This occurs when the non-commutation quantum conditions are imposed on the wave function, making the transition from a quantum *particle* theory to a quantum *field* theory.

Universals and existence: My treatment is indebted to Gracia, *Individuality*, pp. 170–178. Bell prefers to consider "be-ables" rather than "observ-ables" in *Speakable and Unspeakable*, pp. 52–62, 173–180, implying the superiority of being over mere observation. Yet here he does not seem to ask whether being is finally so fundamental.

Planck's constant as a "conversion factor": In Einstein's special theory of relativity, the speed of light is a "conversion factor" between an amount of time t and an amount of space x traversed by light, according to $c = x/t$. (See Edwin F. Taylor and John A. Wheeler, *Spacetime Physics* (San Francisco: W. H. Freeman, 1992) [second edition]). In a similar spirit, one might say that Planck's constant h is a "conversion factor" between the electron's ceaseless internal oscillations (its frequency v) and its externally observable energy E, which are related by Planck's $h = E/v$. See my paper "Identity," pp. 62–63.

Quantum field theory: Among the classic treatments, see J. D. Bjorken and S. D. Drell, *Relativistic Quantum Mechanics* and *Relativistic Quantum Fields* (New York: McGraw-Hill, 1964, 1965) and (among more recent books) Michio Kaku, *Quantum Field Theory: A Modern Introduction* (New York: Oxford University Press, 1993). Four outstanding books are particularly helpful: Silvan S. Schweber, *QED and the Men Who Made It: Dyson, Feynman, Schwinger, and Tomonaga* (Princeton: Princeton University Press, 1994); Paul Teller, *An Interpretive Introduction to Quantum Field Theory* (Princeton: Princeton University Press, 1995); Sunny Auyang, *How Is Quantum Field Theory Possible?* (New York: Oxford University Press, 1995); and Tian Yu Cao, *Conceptual Developments of 20th Century Field Theories* (Cambridge: Cambridge University Press, 1997). These books are extremely helpful in delineating the differences between quantum field theory and the earlier quantum theory of particles. See also Michael Redhead, "A Philosopher Looks at Quantum Field Theory," in *Philosophical Foundations of Quantum Field Theory*, ed. Harvey R. Brown and Rom Harré (Oxford: Oxford University Press, 1988), pp. 9–23, as well as other papers in that collection, including James T. Cushing, "Foundational Problems in and Methodological Lessons from Quantum Field Theory," which takes issue with Redhead's argument about the conceptual novelty of that theory. Although van Fraassen, like Cushing, considers quantum field theory to be "equivalent to a somewhat enriched and elegantly stated theory of [individual] particles" (*Quantum Mechanics*, p. 436), Jeremy Butterfield has offered strong arguments against this view in "Interpretation and Identity in Quantum Theory."

Antiparticles and the spin/statistics theorem: See Richard P. Feynman and Steven Weinberg, *Elementary Particles and the Laws of Physics* (Cambridge: Cambridge University Press, 1987), pp. 1–59 and David Finkelstein and Julio Rubinstein, "Connection between Spin, Statistics, and Kinks," *Journal of Mathematical Physics* 9, 1762–1779 (1968).

The "naturalness" of quantum fields: See Steven Weinberg, *The Quantum Theory of Fields* (Cambridge: Cambridge University Press, 1995), vol. 1, and also the papers in Cao, *Conceptual Foundations of Quantum Field Theory*, especially Weinberg's essay "What Is Quantum Field Theory, and What Did We Think It Was?" on pp. 241–251.

Identicality and quantum field theory: See my essay "Euclidean Hyperspace and Its Physical Significance," *Nuovo Cimento B* 108, 1145–1153 (1993). Treiman argues in *Odd Quantum*, pp. 232–254, that quantum field theory automatically implies identicality. However, his argument *assumes* that the mass-parameter of the quantum field is already a constant, which is an essential part of identicality (equality).

String theory: See Michio Kaku, *Introduction to Superstrings* (New York: Springer-Verlag, 1988); for a popular introduction, see Brian Greene, *The Elegant Universe* (New York: W. W. Norton, 1999). Ironically, quantum field theories admit some stringlike solutions, so that string theory may be finally expressible in terms of quantum field theory, as Weinberg argues in "What Is Quantum Field Theory?"

Philosophic aspects of "reidentification": See Strawson, *Individuals*, pp. 31–38, 55–58.

The two yous: I owe this example to Chisholm, "Problems of Identity," pp. 8–18. See also Marjorie S. Price, "Identity through Time," *Journal of Philosophy* 74, 201–217 (1977) and Graeme Forbes, "Origin and Identity," *Philosophical Studies* 37:4, 353–362 (1980).

Plotinus: *The Enneads*, tr. Stephen MacKenna (Burdett, NY: Larson Publications, 1992), pp. 432 (VI.1.10: principle beyond being), 667 (VI.7.38: identity without the affirmation of being), 426 (V.1.4: identity well pleased).

Epilogue

Crowds: See Élias Canetti, *Crowds and Power* (New York: Farrar Straus Giroux, 1984), pp. 17–22, 29–30.

Participation mystique: Lucien Lévy-Bruhl, *How Natives Think* (New York: Washington Square Press, 1966), p. xviii. Lévy-Bruhl also notes a profound alteration in the meaning of "existence" in what he calls prelogical thought: "The verb 'to be' (which moreover is nonexistent in most of the languages of undeveloped peoples) has not here the ordinary copulative sense it bears in our languages. It signifies something different, and something more. It encompasses both the collective representation and the collective consciousness in a participation that is actually lived, in a kind of symbiosis effected by identity of essence" (p. 75).

Keats and the sparrow: *The Selected Letters of John Keats,* ed. Lionel Trilling (New York: Doubleday, 1956), p. 101.

Individuality as a mode: See Gracia, *Individuality,* pp. 134–140.

Acknowledgments

I salute Larry Cohen and his associates at the MIT Press, who shared my vision for this book and whose wonderful support was essential.

I remember with gratitude my colleagues and fellow students at St. John's College in Santa Fe who have shared my interest in these questions.

Millicent Dillon, Jorge Gracia, and Curtis Wilson have generously shared their comments and suggestions. I am deeply grateful for their encouragement and advice.

Finally, Ssu, Andrei, and Alexei—sharers of all that I am.

Index